Possibility Theory

An Approach to Computerized
Processing of Uncertainty

Possibility Theory

An Approach to Computerized Processing of Uncertainty

Didier Dubois and Henri Prade

CNRS, Languages and Computer Systems (LSI)
University of Toulouse III
Toulouse, France

With the Collaboration of

Henri Farreny
Roger Martin-Clouaire and
Claudette Testemale

Translated by

E. F. Harding
University of Cambridge

in association with First Edition

Plenum Press • New York and London

Library of Congress Cataloging in Publication Data

Dubois, Didier.
   [Théorie des possibilités. English]
   Possibility theory: an approach to computerized processing of uncertain-
ty / Didier Dubois and Henri Prade: with the collaboration of Henri Farreny,
Roger Martin-Clouaire, and Claudette Testemale; translated by E. F. Harding in
association with First Edition.
        p.   cm.
   Translation of: Théorie des possibilités.
   Bibliography: p.
   Includes index.
   ISBN 0-306-42520-3
   1. Fuzzy sets—Data processing. 2. Possibility—Data processing. I. Prade, Henri
M. II. Title.
QA248.D74313   1986
511.3′2—dc19   87-32179
                                                                                CIP

This volume is a translation, with some revisions, of *Théorie des Possibilités:
Applications à la Représentation des Connaissances en Informatique,* published by
Masson (Paris) in 1985. Original Edition © 1985 Masson, Editeur, Paris.

© 1988 Plenum Press, New York
A Division of Plenum Publishing Corporation
233 Spring Street, New York, N.Y. 10013

Printed in the United States of America

Foreword

In the evolution of scientific theories, concern with uncertainty is almost invariably a concomitant of maturation. This is certainly true of the evolution of physics, economics, operations research, communication sciences, and a host of other fields. And it is true of what has been happening more recently in the area of artificial intelligence, most notably in the development of theories relating to the management of uncertainty in knowledge-based systems.

In science, it is traditional to deal with uncertainty through the use of probability theory. In recent years, however, it has become increasingly clear that there are some important facets of uncertainty which do not lend themselves to analysis by classical probability-based methods. One such facet is that of *lexical elasticity*, which relates to the fuzziness of words in natural languages. As a case in point, even a simple relation between *X, Y,* and *Z,* expressed as *if X is small and Y is very large then Z is not very small*, does not lend itself to a simple interpretation within the framework of probability theory by reason of the lexical elasticity of the predicates *small* and *large*.

Historically, the development of possibility theory as a branch of the theory of fuzzy sets was motivated in large measure by the need for a systematic way of dealing with lexical elasticity and other forms of uncertainty which are not probabilistic in nature. Basically, possibility theory—as its name implies—deals with the possible rather than probable values of a variable with possibility being a matter of degree. There are, however, some important connections between possibility and probability which are discussed in an insightful way by Dubois and Prade. One such connection relates to the theory of random sets—a subject to which important contributions have been made by Orlov, Goodman and Nguyen, Wang Peizhuang, and others, and which in turn is related to a theory of probabilistic sets developed by Hirota and extended by Pedrycz and Czogala. Another connection relates to the Dempster–Shafer theory of belief and plausibility, which in turn is related to the theory of random sets. An important aspect of these connections is that, in essence, they stem from the fact that any fuzzy set may be represented as a convex combination of a continuous or discrete family of nonfuzzy sets. In particular, when the family is nested, its elements may be interpreted as the level sets (or cut sets) of a fuzzy set and the convex combination becomes an arithmetic average. These and related observations contribute to a better understanding of the conceptual framework of possibility theory, but what is important to note is that the agenda and techniques of possibility theory are distinct from those of probability theory, the theory of random sets, and the Dempster–Shafer theory of belief and plausibility.

Viewed in this perspective, the distinctness of possibility theory does not imply that it is an alternative to probability theory. Rather, it is in a complementary relation to the latter in the sense that it addresses a class of issues in the management of uncertainty which are not addressed by probability theory. A case in point is a central problem in the combination of evidence in expert systems, namely, how to combine evidence under incomplete information. In this case, possibilistic evidence is, in general, compositional whereas probabilistic evidence is not.

Over the past decade, possibility theory has evolved into a substantive body of concepts and techniques with a variety of applications ranging from pattern classification and control processes to decision analysis and the management of uncertainty in expert systems. The authors of the present volume played a major role in the development of both the theory and its applications, especially through their elegant and highly skillful use of the dual concepts of possibility and necessity. One cannot but be greatly impressed by the thoroughness of their exposition and their success in making the theory easy to understand and simple to implement. In a compelling way, they have produced an outstanding text which in the years ahead is likely to have an important impact on the way in which uncertainty is dealt with in artificial intelligence, operations research, systems analysis, and many other fields. We owe to Dubois and Prade our thanks and congratulations.

Lotfi A. Zadeh

Computer Science Division
University of California
Berkeley, California 94720

Preface to the English Edition

This book is a translated and augmented version of a text originally published in French in mid-1985. Alterations to the original version consist of updating references and improving the text when new results have become available. The latter type of modifications can be found especially in Chapter 1, Section 6 and Chapter 4, Section 2. Moreover, Appendix A to Chapter 4, which describes a practical technique to implement fuzzy reasoning, was hard to follow in the French edition. Its exposition has been both simplified and expanded for the sake of easy reading. The programs at the ends of Chapters 2–6 are those of the French edition, except the inference engine in LISP, at the end of Chapter 4, which has been extensively revised. In order to help the English-speaking reader work through these programs, the comments have been translated into English by the authors.

DIDIER DUBOIS
HENRI PRADE

Preface

The advent of the computer, by raising our expectations of solving new practical problems based on more and more complex models, has brought about the need to gather ever more sophisticated data. It has been observed that much of this information often cannot be obtained as precise and definite numbers, that purely symbolic treatment can be inadequate, and that for various reasons—imperfect measuring instruments, or, as often happens, the fact that the sole source of information is a human being—the information is imprecise, incoherent, and in any case incomplete. Because of this, the elaboration of theories and techniques for representing imprecision and uncertainty, and of methods for handling them, plays a substantial part in the development of computer science and is likely to influence its progress and that of the disciplines that make use of it.

Not only is imprecision inaccessible to measurement, but there are situations in which, even if it could be measured, the results would be of little use or difficult to interpret. Such is the case in analyzing the behavior of a complex or many-dimensional system. An aggregated model sometimes yields more immediately assimilated information than a very detailed, and therefore more exact, model. As Zadeh has stressed, "as the complexity of a system increases, our ability to make precise and yet significant statements about its behavior diminishes until a threshold is reached beyond which precision and significance (or relevance) become almost mutually exclusive characteristics"[1].

This principle of incompatibility is linked with the way humans perceive and reason: we mainly use summary representations of reality, which are therefore imprecise and generally tainted with subjectivity.

In fact a "good model" must achieve a compromise that avoids any excessive precision, likely to be arbitrary and so uncertain. The above remarks are not meant to criticize the scientific process: since Heisenberg's uncertainty principle was discovered, irreducible imprecision has become a familiar phenomenon. At the same time, to allow for imprecision, even for that which arises from the limited capacities of the human understanding, does not rule out that care for rigor which, above even the search for precision, must be intrinsic to the scientific process.

Interval analysis [2] and probability theory are two classical approaches to the representation of imperfect information. They turn out to be inadequate to deal with new needs. Interval analysis lacks nuance and applies only to the treatment of inaccurate numerical data [2]. Probability theory seems to offer too normative a framework to take account of uncertain judgment. This last point is discussed in Chapter 1.

Nevertheless, it is not novel to question the suitability of additive probability as a model of uncertain judgment. In a remarkable paper on the history of science, Shafer [3] recalls that until the end of the 17th century the concepts of chance (associated with randomness) and of probability (an attribute of opinion) were viewed independently of each other. Originally, the doctrine of chances and the calculus of probabilities were developed separately, in the latter case without the axiom of additivity. Shafer [3] shows that a by no means negligible portion of the work of James Bernoulli is concerned with nonadditive probabilities. The success of the additive theory of probability is linked with the rapid expansion of physical science, in which the problems of the modeling of judgment faded into the background. When the latter concerns re-emerged in the 20th century the first work they gave rise to was marked by a partiality for the additivity of degrees of belief, especially in decision theory. However, in the 1950s, the English economist Shackle [4] proposed a nonprobabilistic framework in which the decision process is analyzed in terms of "possibility." In the 1960s, some statisticians, particularly Dempster [5], introduced the notions of upper and lower probabilities, no longer additive, for dealing with incomplete data. Taking up Dempster's model, Shafer [6] interpreted these upper and lower probabilities as degrees of plausibility and belief, respectively, from a decision-theoretic point of view. With progress in computer science and with the development of artificial intelligence, the need to have a theory of uncertain judgment that went beyond the probabilistic framework suddenly became more pressing, as witness expert systems such as MYCIN (Shortliffe and Buchanan [7]), and the increasing interest among the artificial intelligence community in work such as Shafer's.

Possibility theory, formulated by Zadeh in 1977, offers a model for the quantification of judgment which also allows a canonical generalization of the interval analysis. In this framework, the uncertainty of an event is described both by the degree of possibility of the event itself and by the degree of possibility of the contrary event, these two measures being but weakly related. The complement (with respect to 1) of the possibility of the contrary event can be interpreted as the degree of necessity (certainty) of the event itself. Thinking about uncertainty in terms of more or less possible events and more or less certain events appears natural, and seems to be commonly employed by the human mind. In fact, this point of view exactly reflects Shackle's intuition, which thus finds its formalization in possibility theory. Moreover, it can be shown (see Chapters 1 and 4) that the distinction between "possible" and "necessary" is, mathematically, a special case of Shafer's proposed distinction between "plausible" and "credible".

Possibility theory emerged from the notion of fuzzy sets, developed by Zadeh himself in the 1960s. This concept tries to take account of the fact that an object may more or less correspond to a certain category in which one attempts to place it. When the degrees of possibility can only take the black-or-white values 0 or 1, the calculus of possibility exactly coincides with interval analysis, in which the imprecise information is represented in terms of sets of possible values (instead of exact values). In a calculus of degrees of possibility, such sets become fuzzy sets. These remarks emphasize the double relationship between possibility theory and, on the one hand, (naive) set theory and, on the other, the concept of measure (as in measure theory). Furthermore, it is known on the one hand that fuzzy sets can be related to the multivalued logics developed by the Polish school in the 1930s, and on the other hand that (as noted by Kampé de Fériet [8] or Fortet and Kambouzia [9]), fuzzy set membership functions can be interpreted as Shaferian plausibility functions, or as "characteristic functions" of random sets.

One merit of possibility theory is at one and the same time to represent imprecision (in the form of fuzzy sets) and to quantify uncertainty (through the pair of numbers that measure possibility and necessity). But it does so in the most qualitative way possible since the calculations proceed essentially by the use of the "minimum" and "maximum" operators. In a sense, possibility theory takes literally (and contrarily) Rutherford's famous saw: "Qualitative is nothing but poor quantitative"; the point of view in possibility theory is to consider the quantitative as a limiting case of the qualitative.

This book is arranged in six chapters. Basic concepts are presented and discussed in Chapter 1. Here the emphasis is on the relations that hold between measures of possibility and measures of probability. Mathematical tools are the object mainly of Chapters 1 and 2, with further developments in Chapters 3 and 4. Chapters 2–6 illustrate the use of possibility theory in various domains such as operational research, artificial intelligence, and databases, by means of simple examples and corresponding computer programs. The chapters occur in the order in which they are best read, but Chapters 4, 5, and 6 are relatively independent of each other. The programs that are given at the ends of the chapters demonstrate the feasibility of the techniques expounded, but are not necessarily optimized for performance. The choice of BASIC [10] for some of them is justified by the widespread use of this language; the nature of some of the other examples requires the use of LISP [11]. The range of applications described here makes no pretence to completeness. For example, there is nothing in this book about automatic classification, process control, or optimization—all domains to which possibility theory or fuzzy set theory has already been applied with some success. A broader if less up-to-date view can be found in the earlier book by the present authors (reference 6 of Chapter 1).

Chapters 4, 5, and 6 were conceived and written in collaboration with Roger Martin-Clouaire, Henri Farreny, and Claudette Testemale, respectively. The LISP programs that appear in those chapters, were written by them. The BASIC programs are the work of Philippe Chatalic. Claude Tessier kindly provided Figures 3.2–3.13.

References and Notes

1. ZADEH, L. A. (1973). Outline of a new approach to the analysis of complex systems and decision processes. *IEEE Trans. Syst. Man Cybern.*, **3**, 28–44.
2. See references 16 and 17 of Chapter 2.
3. SHAFER, G. (1978). Non-additive probabilities in the works of Bernoulli and Lambert. *Arch. Hist. Exact Sci.*, **19**, 309–370.
4. SHACKLE, G. L. S. (1961). *Decision, Order and Time in Human Affairs*. Cambridge University Press, Cambridge, 2nd edition.
5. See reference 1 of Chapter 1.
6. See reference 28 of Chapter 1.
7. See reference 40 of Chapter 4.
8. KAMPÉ DE FÉRIET, J. (1980). Une interprétation des measures de plausibilité et de crédibilité au sens de G. Schafer et de la fonction d'appartenance définissant un ensemble flou de L. Zadeh. *Publ. IRMA* (Lille), **2**(6, 2), II-01–II-22.
9. FORTET, R., and KAMBOUZIA, M. (1976). Ensembles aléatoires et ensembles flous. *Publ. Econométriques*, **IX**(1), 1–23.
10. ALPHA-BASIC on the AMS.
11. LE-LISP under VMS on the VAX 750 for Chapter 4, LE-LISP on the Macintosh (512K) for Chapter 5, and MACLISP under MULTICS on the DPS8 for Chapter 6.

Contents

1

Measures of Possibility and Fuzzy Sets

The material in this book is based on a nontraditional approach to the imprecise and the uncertain. The basic concept is the measure of possibility. The object of this introduction is to provide motivation and context, to define measures of possibility, and to present basic notions necessary for understanding the later chapters. It appeals considerably to results contained in the authors' theses [3, 24], among other references.

1.1. Imprecision and Uncertainty

Imprecision and uncertainty can be considered as two complementary aspects of a single reality, that of imperfect information. We shall assume here that an item of information can be expressed as a logical proposition with predicates, and possibly quantifiers. A corpus of knowledge will be a collection of items of information possessed by an individual (or a computer system, or a group of individuals), relating to a single problem. The predicates that occur in the expressions of this information can then be interpreted as subsets of a single domain of reference. A proposition can also be considered as an affirmation concerning the occurrence of an event. Such events can themselves be represented as subsets of this reference domain, which can therefore be called "the sure event." Thus we have three equivalent ways to envisage a collection of items of information, depending on whether we emphasize its structure (logical aspect), its content (set-theoretic aspect), or the relation of the items to real events (factual aspect).

1

From the practical point of view, an item of information will be defined as a quadruple (attribute, object, value, confidence). "Attribute" refers to a function that attaches a value (or a set of values) to the object whose name figures in the item of information. This value corresponds to a predicate, that is to say, to a subset of the reference domain associated with the attribute. The confidence is an indication of the reliability of the item of information. Clearly the four entities making up the item of information can be composite (several objects, many attributes, n-ary predicate, different degrees of confidence). Moreover, variables can be introduced, especially for the objects, if the item of information involves quantifiers.

In this context we can clearly distinguish the concepts of imprecision and uncertainty: imprecision relates to the content of an item of information (the "value" component of the quadruple) while uncertainty relates to its truth, understood as its conformity to a reality (the "confidence" component of the quadruple).

The uncertainty of an item of information can be assessed by means of qualifiers such as "probable," "possible," "necessary," "plausible," or "credible," to which we shall attempt to give a precise meaning. The modality of being *probable* has been extensively studied for two centuries. It has two distinct connotations, one of which is physical, tied to statistical experiments, and concerned with the frequency of occurrence of an event. The other is epistemic: here "probable" refers to a subjective judgment. The modalities of *possible* and *necessary* go back to Aristotle, who stressed their duality (if an event is necessary, then its contrary is impossible). Oddly, and in contrast with the concept of "probable," the "possible" and the "necessary" are often considered as all-or-nothing categories. But, like "probable," "possible" has two interpretations: physical (as a measure of the material difficulty of performing an action), and epistemic (as a subjective judgment that does not much commit its maker). "Necessary," on the other hand, is a much stronger notion, in either the physical or epistemic sense (subjective necessity amounts to certainty). It is natural to admit degrees of possibility and of necessity, as for probability (a nuance already present in everyday language, where one says "very possible" for example). The connotations of "plausible" and "credible" are specifically epistemic and refer respectively to "possible" and 'necessary." Each corresponds to a mode of inference based on a given corpus of knowledge: anything that can be deduced from the corpus is *credible*; anything that does not contradict the corpus is *plausible* (inductive aspect). As to "likely," usage seems to give it a sense near to "probable," while "plausible" seems closer to "conceivable."

Here are some examples of uncertain propositions:

- It is probable that John is at least 1.70 m tall.

$$\triangleq \text{(height, John, } \geqslant 1.7 \text{ m, probable)}$$

- The probability of 10 mm of rain tomorrow is 0.5.

$$\triangleq \text{(quantity, rain tomorrow, 10 mm, probability } = 0.5)$$

An item of information will be called *precise* when the subset corresponding to its "value" component cannot be subdivided. Depending on what aspect of the information is being emphasized, we shall speak of an *elementary proposition* (i.e., implied by no other proposition save the necessarily false proposition), of a *singleton* (set-theoretic aspect), or of an *elementary event*. The property of being "precise" depends of course on the definition of the domain of reference (on its "granularity"—e.g., the choice of a unit of measurement). In other cases, we shall speak of *imprecise* information.

In English, there are other qualifiers that refer to imprecision, for example, "vague," "fuzzy," "general," or "ambiguous." *Ambiguity* is a kind of imprecision allied to language, sometimes due to homonymy: an item of information is ambiguous to the extent that it refers to several possible contexts or reference sets. Imprecision of this kind will not be considered in this text: we shall suppose that the domain of reference associated with an element of information is known. *Generality* is a (beneficial) form of imprecision allied to the process of abstraction; an item of information is general if it designates a class of objects of which it expresses a common property. But *vagueness* or *fuzziness* in an item of information resides in the absence of a clear boundary to the set of values attached to the objects it refers to. Many qualifiers in everyday language are fuzzy, and reflect the features of generality. The following are examples:

- An imprecise, nonfuzzy proposition:

$$(x = y \text{ to within } \varepsilon) \triangleq \text{(equality, } (x, y), \text{ to within } \varepsilon, 1)$$

- An imprecise fuzzy proposition:

$$(x \text{ approximately } = y) \triangleq \text{(equality, } (x, y), \text{ approximately, } 1).$$

The vague term "approximately" designates several more or less adequate values of ε.

It goes without saying that an item of information can be both vague and uncertain at the same time, as witnessed by the following:

- It is probable that it will rain *a lot* tomorrow.

$$\triangleq \text{(quantity, rain tomorrow, a lot, probable)}$$

Given a set of items of information, the opposition between precision and uncertainty is expressed by the fact that in rendering the content of a proposition more precise, one will tend to increase its uncertainty. Conversely, the uncertain character of an item of information will in general confer a certain imprecision on the conclusions to be drawn from it.

1.2. Traditional Models of Imprecision and Uncertainty

Traditionally, two methods of representing imperfect information are available: probability theory and what is known as interval analysis. Let us briefly look at the domains where these are used.

Probability theory today is a tried and tested mathematical theory: its axioms are clear and not disputed, and it has attained a very advanced development. The fundamental axiom is that of additivity of the probabilities of disjoint events. The controversial aspects of probability theory bear on its interpretation: what reality is supposed to be represented by this mathematical model? Historically, the theory was used for the "calculus of chances" in games of chance, the probability of an event being defined as the ratio of the number of favorable cases to the total number of possible cases. The rather nonoperational character of this definition gave rise to the "frequentist school," for whom probabilities are to be interpreted as limits of frequencies of observed events. The third school, the so-called "subjectivists," tried to avoid the difficulties the frequentists encountered in applying the theory (sufficiently many observations, repeatability of experiments, and so on) by proposing probability as a measure of the feeling of uncertainty. The numerical value of a probability is, then, interpreted as proportional to the sum an individual would be willing to pay should a proposition that he asserts prove false. In so far as this sum exists, it is shown that the measure of uncertainty so defined obeys the axioms of probability theory, provided that the behavior of the individual satisfies conditions of "rationality" (Savage [27]). On this basis, the subjectivists succeeded in showing that Kolmogorov's axioms were the only reasonable basis for evaluating subjective uncertainty.

This extreme attitude can be contested from a philosophical and from a practical point of view. First of all, it seems difficult to maintain that every uncertain judgment obeys the rules of betting. The monetary commitment that forms part of the model could prevent an individual from uncovering the true state of his knowledge, for fear of financial loss. Thus, a professional gambler will distribute his stakes evenly if he knows that all the options on which he is betting have equal strength. In the absence of any information, the neophyte will do the same, because it is the most prudent strategy. Subjective probabilities allow no distinction between these two states of knowledge and seem ill adapted to situations where this knowledge is sparse. In particular, the limiting case of total ignorance is very poorly handled by the probabilistic model, which presupposes that a set of mutually disjoint possible events has been identified, to which are assigned, by virtue of the principle of maximum entropy, equal probabilities in the finite case. In the case of total ignorance, it seems to be ruled out that one is capable of identifying all these events, and therefore disputable that the measures of uncertainty attributed to them should depend on the number of alternatives, as is the case with probabilities.

From the practical point of view, it is clear that the numbers given by individuals to describe, in terms of probabilities, for example, the state of their knowledge must be considered for what they are, namely, approximate indications. Subjective probability theory does not seem to be concerned with this type of imprecision, considering that a rational individual must be able to furnish precise numbers, when proper procedures for their elicitation are used.

To sum up, probability theory seems to be too normative a framework to take all the aspects of uncertain judgment into account.

Interval analysis, commonly applied in physics, merely serves to throw the inaccuracies of measuring instruments, in the form of intervals, back onto the magnitudes estimated by the measurements. In mathematical terms, one evaluates the image of a function whose arguments are subsets. Interval analysis has no gradations: while one does not know the exact value of a parameter, one does know the exact limits of its domain of variation. Note that, given an imprecise measurement M of the magnitude X, propositions of the type "X belongs to the interval I" can naturally be qualified by the modalities of "possible" and "necessary" since

1. If $M \cap I$ is not empty then "$X \in I$" is possibly true.
2. If $M \subseteq I$ then "$X \in I$" is necessarily true.

Here we emphasize the relationships between these modalities and the theory of sets: the *possible* is to be evaluated with respect to the set-theoretic intersection between the contents M and I of the two propositions "$X \in M$" and "$X \in I$"; the *necessary* is evaluated with respect to set-theoretic inclusion.

The all-or-nothing nature of interval analysis, in contrast to probability theory, which admits gradations, introduces an asymmetry between them, which one would like to remove. It is clear that the latter does not generalize the former, since a function of a uniformly distributed random variable (the probabilistic counterpart of an error interval) does not in general itself have a uniform distribution. One of the major contributions of this book is to propose a canonical generalization of interval analysis that admits of appropriate gradations.

It commonly occurs that imprecision of the error-of-measurement kind is present at the heart of a series of trials intended to exhibit a random phenomenon. In such a case it can be observed that one can hardly represent the information in a purely probabilistic form without introducing further hypotheses. In fact, a hypothesis fundamental to the applicability of probability to statistics is that there should be a bijection between the sample space and the event space: to every event is associated the set of sample points that realize it (which is nonempty if the event is not impossible), and for every pair of distinct events there is at least one sample point that realizes one but not the other. This hypothesis therefore allows the sure event to be partitioned into elementary events, each corresponding to a specific sample point. In the case of a collection of statistical data, this amounts to supposing that there is a nontrivial partition of the set of realizations such that the result of each experiment can be associated with one and only one element of this partition, i.e., is an elementary event.

We can find situations where this hypothesis for the partitioning of trials does not hold. For example, if the experiments yield error intervals, one has in general little possibility of associating them with disjoint classes of realizations. The physicist, rather, expects to encounter the converse situation—namely, sufficient coherence between the intervals resulting from independent measurements to enable him, by cross-checking, to diminish the measurement error. One can see that even with "objective" repeated phenomona, probability theory cannot always be directly used. A probabilistic model is suitable for the expression of precise but dispersed information. Once the precision is lacking, one tends to quit the domain of validity of the model.

This short discussion of the limitations of traditional models of imprecision and uncertainty is intended to justify the search for a wider framework, embracing the two concepts, in which probability and interval analysis can take their proper places and in which their similarities and their differences can be emphasized. This general framework, which is merely sketched in the present book, will naturally include a new family of measures of uncertainty closely allied to interval analysis: the measures of possibility. These set functions are framed in total contrast to measures of probability. While the latter can, as described above, be applied to results that are precise but contradictory, measures of possibility are the natural tool for summarizing a corpus of results that are imprecise, but coherent.

1.3. Confidence Measures

Let us consider a set of events associated with a corpus of imprecise and uncertain knowledge and considered as subsets of a reference set Ω called the "(always) sure event." The empty set is identified with the "(always) impossible event." It is supposed that with an event $A \subseteq \Omega$ is associated a real number $g(A)$ furnished by an individual who is in possession of the corpus (or by a data-processing procedure applied to information stored in the memory of a computer system). $g(A)$ measures the confidence one may have in the occurrence of the event A, taking the state of knowledge into account. By convention, $g(A)$ increases with increasing confidence. Moreover, if A is a sure event then one takes $g(A) = 1$, and if A is an impossible event, then $g(A) = 0$, In particular

$$g(\varnothing) = 0 \quad \text{and} \quad g(\Omega) = 1 \tag{1.1}$$

Nevertheless, $g(A) = 1$ (or 0) does not necessarily mean that A is sure (or impossible).

The weakest axiom that one could conceive to ensure that the set function g has a minimum of coherence is that it should be monotonic with respect to inclusion:

$$A \subseteq B \Rightarrow g(A) \leqslant g(B) \tag{1.2}$$

This axiom means that if the even A implies another event B, then one always has at least as much confidence in the occurrence of B as in the occurrence of A.

Such set functions were proposed by Sugeno [29] for the evaluation of uncertainty, under the name of "fuzzy measures." A. Kaufman suggested the name "valuations." We here adopt the term "confidence measures." It is appropriate to point out that they are not measures in the usual additive sense, unless this is expressly indicated.

When Ω is an infinite reference set, one can introduce axioms of continuity written as follows: For every nested sequence $(A_n)_n$ of sets $A_0 \subseteq A_1 \subseteq \cdots A_n \subseteq \cdots$, or $A_0 \supseteq A_1 \supseteq \cdots A_n \supseteq \cdots$, we have

$$\lim_{n \to +\infty} g(A_n) = g\left(\lim_{n \to +\infty} A_n\right) \tag{1.3}$$

A confidence measure will be supposed to satisfy (1.3) for at least one of the two kinds of sequence (increasing or decreasing).

1.3.1. Measures of Possibility and of Necessity

The following inequalities are immediate consequences of the monotonicity axiom (1.2), and concern disjunctions $(A \cup B)$ and conjunctions $(A \cap B)$ of events:

$$\forall A, B \subseteq \Omega, \qquad g(A \cup B) \geqslant \max(g(A), g(B)) \tag{1.4}$$

$$g(A \cap B) \leqslant \min(g(A), g(B)) \tag{1.5}$$

We thus naturally find, as limiting cases of confidence measures, functions, denoted by Π, such that

$$\forall A, B, \qquad \Pi(A \cup B) = \max(\Pi(A), \Pi(B)) \tag{1.6}$$

and called *possibility measures* by Zadeh [37]. One may be surprised that in (1.6) it is not supposed that the sets A and B are disjoint. It is easy to verify that if (1.6) is true for every pair of disjoint sets $(A \cap B = \varnothing)$ then it is true for every pair of events (Dubois and Prade [6]). The term "possibility" for these confidence measures can be justified from several points of view:

- Suppose $E \subseteq \Omega$ is an event considered as sure. A function Π, taking values in $\{0, 1\}$ and satisfying (1.6), is easily defined by

$$\Pi(A) = 1 \qquad \text{if } A \cap E \neq \varnothing$$
$$= 0 \qquad \text{otherwise} \tag{1.7}$$

It is clear what $\Pi(A) = 1$ means in this context: A is possible. This suggests the relationship that possibility measures may have with interval analysis (see above). In particular, if A and $\bar{A}$ are two contradictory events ($\bar{A}$ is the complement of A in Ω), then (1.6) implies

$$\max(\Pi(A), \Pi(\bar{A})) = 1 \qquad (1.8)$$

which can be interpreted as the fact that of two contradictory events, one at least is completely possible; further, when one event is judged to be possible, this does not prevent the contrary event from being so as well, which is consistent with the semantics of judgments of possibility that commit their makers very little.

• Finally, axiom (1.6) is consistent with the physical notion of possibility: in order to realize $A \cup B$ it is sufficient to realize the easier (or less costly) of the two.

When the set Ω is finite, every possibility measure Π can be defined in terms of its values on the singletons of Ω:

$$\forall A, \qquad \Pi(A) = \sup\{\pi(\omega) \mid \omega \in A\} \qquad (1.9)$$

where $\pi(\omega) = \Pi(\{\omega\})$; π is a mapping of Ω into $[0, 1]$ called a *possibility distribution*. It is *normalized* in the sense that

$$\exists \, \omega, \qquad \pi(\omega) = 1 \qquad (1.10)$$

since $\Pi(\Omega) = 1$.

N.B. Equation (1.9) holds even if, like Zadeh [37], one does not require $\Pi(\Omega) = 1$. In this case (1.8) and (1.10) hold if 1 is changed to $\Pi(\Omega)$. □

When the set Ω is infinite, a possibility distribution does not necessarily exist, and indeed does so only if axiom (1.6) is extended to infinite unions of events (Nguyen [21]). In practice, one can always start with a possibility distribution and construct Π by using (1.9). In the most general case, possibility measures do not satisfy the continuity axiom (1.3) for decreasing nested sequences of sets (Puri and Ralescu [25]).

The other limiting case of confidence measures is obtained by imposing equality in (1.5). One then obtains a class of set functions, called *necessity measures* and denoted by N, which satisfy the axiom dual to (1.6):

$$\forall A, B, \qquad N(A \cap B) = \min(N(A), N(B)) \qquad (1.11)$$

A function N with values in $\{0, 1\}$ can easily be constructed in terms of knowledge of a sure event E on putting

$$N(A) = 1 \quad \text{if } E \subseteq A$$
$$= 0 \quad \text{otherwise} \tag{1.12}$$

$N(A) = 1$ clearly means that A is sure (necessarily true). Further, it can easily be seen that a set function N satisfies (1.11) if and only if the function Π defined by

$$\forall A, \quad \Pi(A) = 1 - N(\bar{A}) \tag{1.13}$$

is a possibility measure. Equations (1.12) and (1.13) justify the name "necessity measure" for N. Equation (1.13) is a numerical expression of a duality relationship between the modalities of the possible and the necessary (in modal logic), which postulates that an event is *necessary* when its contrary is impossible. This duality relationship means that one can always construct a necessity function from a possibility distribution by means of

$$N(A) = \inf\{1 - \pi(\omega) \mid \omega \notin A\} \tag{1.14}$$

Necessity measures satisfy the relation

$$\min(N(A), N(\bar{A})) = 0 \tag{1.15}$$

which prohibits two contrary events from both being the slightest bit necessary at the same time. From (1.13) and (1.15) [or (1.8)] it is easily verified that

$$\forall A \subseteq \Omega, \quad \Pi(A) \geqslant N(A) \tag{1.16}$$

which agrees with the intuition that an event becomes possible before becoming necessary. In addition, there are the following relations which are stronger than (1.16):

$$N(A) > 0 \Rightarrow \Pi(A) = 1 \tag{1.17}$$
$$\Pi(A) < 1 \Rightarrow N(A) = 0 \tag{1.18}$$

1.3.2. Possibility and Probability

When the knowledge one has of the occurrence of events is given in the form of observed frequencies of elementary events, the confidence measure P naturally satisfies the additivity axiom

$$\forall A, \forall B, \text{ with } A \cap B = \varnothing, \qquad P(A \cup B) = P(A) + P(B) \quad (1.19)$$

so that it is a probability measure which, of course, is monotonic in the sense of (1.2). Axiom (1.19) is the probabilistic equivalent of axioms (1.6) and (1.11).

In the finite case, the equivalent of (1.9) and (1.14) is

$$P(A) = \sum_{\omega \in A} p(\omega) \qquad (1.20)$$

where $p(\omega) = P(\{\omega\})$. The normalization condition $\sum_{\omega \in \Omega} p(\omega) = 1$ is the analog of (1.10). Measures of probability, possibility, and necessity have in common that all three can be characterized by a distribution on the elements of the reference set.

Finally, the counterpart of (1.8) and (1.15) is the well-known relation

$$P(A) + P(\bar{A}) = 1 \qquad (1.21)$$

while (1.8) and (1.15) only imply that

$$N(A) + N(\bar{A}) \leqslant 1 \qquad (1.22)$$

$$\Pi(A) + \Pi(\bar{A}) \geqslant 1 \qquad (1.23)$$

Here can be seen one of the major differences between possibility and probability: the probability of an event completely determines the probability of the contrary event. The possibility (or the necessity) of an event, and that of the contrary event, are but weakly linked; in particular, in order to characterize the uncertainty of an event A one needs both of the numbers $\Pi(A)$ and $N(A)$ [subject to (1.17) or (1.18)].

In modeling uncertain judgment (belief), it seems natural not to wish to rigidify the relationship between the indications one has in favor of an event (the degree of necessity) and those that weigh against it (one minus the degree of possibility). In this situation the notion of probability seems less flexible than that of possibility.

Even if the requirement of additivity is kept, it is still possible to construct measures of possibility and necessity, provided one no longer insists that the probability masses are carried by the elementary events. More exactly, let $E_1, E_2, \ldots, E_p$ be subsets, distinct by pairs (but not necessarily disjoint), of Ω (presumed finite), carrying respectively probability masses $m(E_1), \ldots, m(E_p)$ which are such that

$$\sum_{i=1}^{p} m(E_i) = 1 \qquad (1.24)$$

and

$$\forall i, \qquad m(E_i) > 0 \qquad (1.25)$$

The mass $m(E_i)$ can be interpreted as a global allocation of probability to the whole set of elementary events making up E_i, without specifying how this mass is distributed over the elementary events themselves. The events E_i are called "focal elements" (Shafer [28]) and may be used to model imprecise observations. In this situation, the probability of an event A will be imprecise, that is, will be contained in an interval $[P_*(A), P^*(A)]$, defined by

$$P_*(A) = \sum_{E_i \subseteq A} m(E_i) \qquad (1.26)$$

$$P^*(A) = \sum_{E_i \cap A \neq \varnothing} m(E_i) \qquad (1.27)$$

$P_*(A)$ is calculated by considering all the focal elements which make the occurrence of A necessary (i.e., which imply A). $P^*(A)$ is obtained by considering all the focal elements which make the occurrence of A possible. Note that, again, there is a duality relation between P^* and P_*:

$$\forall A, \qquad P^*(A) = 1 - P_*(\bar{A}) \qquad (1.28)$$

It can be verified (Shafer [28]) that the functions P^* and P_* satisfy, respectively, axioms (1.6) and (1.11), that is to say they are, respectively, a possibility (or necessity) measure if and only if the focal elements form a nested sequence of sets. More particularly, if $E_1 \subset E_2 \subset \cdots \subset E_p$, then the possibility distribution π associated with P^* and P_* is defined by (Dubois and Prade [8])

$$\forall \omega, \pi(\omega) = P^*(\{\omega\}) = \sum_{j=i}^{p} m(E_j) \qquad \text{if } \omega \in E_i, \qquad \omega \notin E_{i-1}$$

$$= 0 \qquad \qquad \text{if } \omega \in \Omega - E_p \qquad (1.29)$$

If on the other hand the focal elements are elementary (therefore disjoint), then clearly $\forall A$, $P_*(A) = P^*(A) = P(A)$, so that we come back to a probability measure.

If then a corpus of knowledge is schematized by a set of focal elements (which are the "value" components in the quadruples representing the items of information), it can be clearly seen that probability measures apply to precise but differentiated items of information, while possibility measures reflect imprecise but coherent items (i.e., which mutually confirm each other). Let us note that possibility functions are in this sense more natural for the representation of subjective uncertainty: we do not expect that a single individual will provide us with very precise data, but we would expect the greatest possible coherence in his statements. On the other hand, precise but variable data are the usual results of carefully observing a physical phenomenon.

Of course, a corpus of information will contain data that in general are neither precise nor completely coherent. Probability, and possibility–necessity, correspond to two extreme and therefore idealized situations.

The formulas (1.26) and (1.27) allow a possibility distribution to be considered as implicitly defining a class of probability measures $\mathcal{P}$ such that

$$\mathcal{P} = \{P \mid \forall A, N(A) \leqslant P(A) \leqslant \Pi(A)\} \tag{1.30}$$

This remark allows the rigorous definition of the notion of mathematical expectation for possibility measures. More precisely, if f is a function defined on Ω, taking values in the reals ($\mathbb{R}$), the upper and lower mathematical expectations $E^*(f)$ and $E_*(f)$, respectively, are defined as the Lebesgue–Stieltjes integrals (Dempster [1]):

$$E_*(f) = \int r \, d\Pi(\{\omega \mid f(\omega) \leqslant r\}) \tag{1.31}$$

$$E^*(f) = \int r \, dN(\{\omega \mid f(\omega) \leqslant r\}) \tag{1.32}$$

The terms "upper" and "lower" are justified by the identities

$$E^*(f) = \sup\{E(f) \mid P \in \mathcal{P}\}; \qquad E_*(f) = \inf\{E(f) \mid P \in \mathcal{P}\} \tag{1.33}$$

These results were obtained by Dempster [1] for the case of finite Ω; see Huber [15] for a more general case.

1.4. Fuzzy Sets

The concept of fuzzy set can be defined without making any reference to measures of uncertainty by changing the usual definition of the characteristic function of a set, so as to introduce *degrees* of membership. This is the logical point of view. Nonetheless, having a measure of uncertainty amounts to a tendency to localize the value of a variable *x* by giving, for each subset *A* of a reference set, the knowledge one possesses about the relation $x \in A$. The family of pertinent subsets for representing the variable *x* will induce a generalized characteristic function of a fuzzy set, the two representations being strictly equivalent for possibility measures.

In the former point of view, a fuzzy set *F* is equivalent to giving a reference set Ω and a mapping, μ_F, of Ω into $[0, 1]$, the unit interval (Zadeh [31]). $\mu_F(\omega)$, for $\omega \in \Omega$, is interpreted as the degree of membership of ω in the fuzzy set *F*. This is the most direct definition. It gives a direct model for the vague categories of natural language ("tall" for example), defined on an objective base which could be a numerical scale (Ω = set of heights) or the set of objects qualified by these categories (Ω = set of persons). $\mu_F(\omega)$ then expresses how much the value (or the object) ω is compatible with the concept *F*. If $\Omega = \mathbb{R}$ (the reals) then *F* is a *fuzzy quantity*.

It is natural to look for standard set-theoretic representations of *F*. When $\mu_F(\omega) \in \{0, 1\} \; \forall \; \omega$, *F* is the same as an ordinary subset of Ω. In this case *F* is called a "crisp subset" of Ω ("crisp" is the usual term, rather than "sharp"). In the contrary case we can choose a threshold $\alpha \in \;]0, 1]$ and define the set

$$F_\alpha = \{\omega \in \Omega \mid \mu_F(\omega) \geq \alpha\} \tag{1.34}$$

F is called a "α-level cut" or "α-cut." F_α contains all the elements of Ω that are compatible with *A* at level at least α. The family $C(F) = \{F_\alpha \mid \alpha \in \;]0, 1]\}$ of α-cuts is monotone:

$$0 < \alpha \leq \beta \leq 1 \Rightarrow F_\alpha \supseteq F_\beta \tag{1.35}$$

It constitutes a representation of *F* by means of classical sets since we have (Zadeh, [33])

$$\forall \; \omega, \qquad \mu_F(\omega) = \sup\{\alpha \mid \omega \in F_\alpha\} \tag{1.36}$$

Conversely, a finite monotone family of sets $\{F_{\alpha_1}, \ldots, F_{\alpha_m}\}$, which have been given weights α_i satisfying (1.35), forms the set of α-cuts of a fuzzy set defined by (1.36). In the case of an infinite family, the condition (1.35) is no longer sufficient; it is also required (Ralescu [26]) that, for every increasing sequence $(\alpha_n)_n$ of elements of $]0, 1]$,

$$\lim_{n \to +\infty} \alpha_n = \alpha \Rightarrow F_\alpha = \bigcap_{n=1}^{\infty} F_{\alpha_n} \tag{1.37}$$

One can also choose to represent F by its *strong* α-cuts, defined by

$$F_{\tilde{\alpha}} = \{\omega \in \Omega \mid \mu_F(\omega) > \alpha\}, \qquad \alpha \in [0, 1[\tag{1.38}$$

Strong α-cuts satisfy (1.35) and (1.36) just like α-cuts. Among the ordinary subsets representing F, two in particular will often be used in what follows:

- The cut of F at level 1, called the *core* or *peak*, denoted $\dot{F}$:

$$\dot{F} = \{\omega \in \Omega \mid \mu_F(\omega) = 1\}$$

- The strong cut of F at level 0, called the *support*, denoted $S(F)$:

$$S(F) = \{\omega \in \Omega \mid \mu_F(\omega) > 0\}$$

N.B. In certain circumstances one may wish to introduce sets other than the α-cuts into the representation of F (cf. Dubois and Prade [7]).

□

The second point of view on a fuzzy set is to consider it as the "trace" of a possibility measure on the singletons of Ω. In fact, a set $E \subset \Omega$ induces a possibility measure Π_E such that $\Pi_E(A) = 1$ if and only if $E \cap A \neq \varnothing$, and $\Pi_E = 0$ otherwise. When a possibility measure has values in the unit interval, one can therefore interpret its distribution π as the membership function of a fuzzy set F, which is the sure event that Π focuses on. In fact, denoting by $[0, 1]^\Omega$ the set of fuzzy subsets of Ω,

$$\forall \Pi, \quad \exists F \in [0, 1]^\Omega, \quad \forall \omega \in \Omega, \quad \Pi(\{\omega\}) = \pi(\omega) = \mu_F(\omega) \tag{1.39}$$

Conversely, from a given fuzzy set a possibility function can be obtained, provided the fuzzy set is *normalized*, that is

$$\exists \omega, \quad \mu_F(\omega) = 1 \tag{1.40}$$

But if we do not require that $\Pi(\Omega) = 1$, then, from (1.9),

$$\forall F \in [0, 1]^\Omega, \qquad \exists \, \Pi, \forall \, \omega \ni \Omega, \qquad \Pi(\{\omega\}) = \pi(\omega) = \mu_F(\omega) \quad (1.41)$$

$\Pi(\Omega) = \sup \mu_F$ is sometimes called the *height* of the fuzzy set F. It is easily seen that if a possibility function is defined by a probabilistic weighting m, then the focal elements form the family of α-cuts of a fuzzy set. Suppose $\{A_1 \subseteq \cdots \subseteq A_p\}$ are these focal elements; then

$$A_i = F_{\alpha_i} \quad \text{or} \quad \alpha_i = \sum_{j=i}^{p} m(A_j)$$

In other words,

$$\forall \, \omega, \qquad \mu_F(\omega) = \sum_{\omega \in F_{\alpha_i}} m(F_{\alpha_i}) \qquad (1.42)$$

This is a "probabilistic" way of representing a fuzzy set, i.e., the "probabilistic" counterpart of (1.36).

N.B. When (1.39) is applied to a probability measure on a finite set, this amounts to considering the atoms of probability as values of the membership function; noting that a probability function represents the idea of a *point* of unknown position and *not of a set* [for if $P(A) \in \{0, 1\}$ then P is a Dirac measure supported on a singleton], this identification may appear to be an abuse of terminology. Viewing a possibility measure as a fuzzy set, a probability measure can only be viewed as a "fuzzy point"! Moreover, contrary to the case of possibility measure, knowledge of $\{P(\{\omega\}) \mid \omega \in \Omega\}$ does not necessarily determine the probability measure, since it is possible (when Ω is infinite) that $P(\{\omega\}) = 0, \forall \, \omega$. $\qquad\qquad \square$

One final point to be discussed concerns whether or not to normalize a fuzzy set. Everything depends on the exhaustive character of the reference set. If, for example, one wishes to represent the set of integers very close to 3.5, it is natural not to normalize the membership function, since the numbers closest to 3.5 are outside $\mathbb{N}$. On the other hand, a fuzzy quantity defined on $\mathbb{R} \cup \{-\infty\} \cup \{+\infty\}$ will in general be normalized, since the reference set is exhaustive. Not normalizing a possibility measure can also be interpreted as a lack of confidence in the information (for instance if it was obtained from conflicting sources), or as due to the fact that the variable associated with the measure may fail to take a value (if for example Ω is the set of realizations of a process which one may perhaps not carry out).

1.5. Elementary Fuzzy Set Operations

The notions of inclusion and equality extend to fuzzy sets: the definitions most widely accepted are due to Zadeh [31]:

Inclusion: $F \subseteq G$ $\quad \forall\, \omega,\ \mu_F(\omega) \leqslant \mu_G(\omega)$ $\hfill (1.43)$

Equality: $F = G$ $\quad \forall\, \omega,\ \mu_F(\omega) = \mu_G(\omega)$ $\hfill (1.44)$

The principal set-theoretic operations (complementation, intersection, and union) have been defined by Zadeh for fuzzy sets as follows:

Complementation: The fuzzy set $\bar{F}$, the complement of F in Ω, is defined by

$$\forall\, \omega, \qquad \mu_{\bar{F}}(\omega) = 1 - \mu_F(\omega) \qquad (1.45)$$

Intersection: The intersection $F \cap G$ of two fuzzy sets F and G on Ω is defined by

$$\forall\, \omega, \qquad \mu_{F \cap G}(\omega) = \min(\mu_F(\omega), \mu_G(\omega)) \qquad (1.46)$$

Union: The union $F \cup G$ of two fuzzy sets F and G on Ω is defined by

$$\forall\, \omega, \qquad \mu_{F \cup G}(\omega) = \max(\mu_F(\omega), \mu_G(\omega)) \qquad (1.47)$$

All these definitions seem to have some arbitrariness, though they do not offend intuition. They agree with the classical definitions when the sets under consideration are ordinary subsets. But in fact, the extensions to fuzzy sets of the operations of complementation, intersection, and union, and of the relations of inclusion and equality, are not unique. The reader is referred to Chapter 3 for a panoply of fuzzy set-theoretic operations different from (1.45)–(1.47). Similarly, (1.43)–(1.44) seem too rigid for comparing, in practice, two fuzzy sets; their interest is largely mathematical. Indices of comparison will be considered below.

Nonetheless, the definitions proposed above can be justified by the richness of the structure that they induce on $[0, 1]^{\Omega}$.

Inclusion as defined by (1.43) is reflexive and transitive.

Complementation (1.45) satisfies the involution $\bar{\bar{F}} = F$, and is in fact unique if it is required that for each pair $(\omega_1, \omega_2) \in \Omega$, turning ω_1 into ω_2 changes the degree of membership of F by an amount equal and opposite to the change in degree of membership of $\bar{F}$, that is to say

$$\forall (\omega_1, \omega_2), \qquad \mu_F(\omega_1) - \mu_F(\omega_2) = \mu_{\bar{F}}(\omega_2) - \mu_{\bar{F}}(\omega_1) \qquad (1.48)$$

The set $[0, 1]^\Omega$ of fuzzy sets on Ω, furnished with the operations defined by (1.45)–(1.47), has a vector lattice structure, that is, all the classical properties of set-theoretic operations are preserved, save the laws of noncontradiction and of the excluded middle, of which only weak versions remain:

$$\forall \omega, \qquad \mu_{F \cap \bar{F}}(\omega) = \min(\mu_F(\omega), 1 - \mu_F(\omega)) \leq 0.5 \qquad (1.49)$$

$$\forall \omega, \qquad \mu_{F \cup \bar{F}}(\omega) = \max(\mu_F(\omega), 1 - \mu_F(\omega)) \geq 0.5 \qquad (1.50)$$

(1.46) and (1.47), that is the "min" and "max" operations, furnish the only possible definitions of fuzzy intersection and fuzzy union that preserve such a structure on the fuzzy sets of Ω. We then obtain an "optimal" structure since it is impossible to preserve the full Boolean lattice structure on $[0, 1]^\Omega$. In particular, the laws of noncontradiction $(F \cap \bar{F} = \varnothing)$ and of the excluded middle $(F \cup \bar{F} = \Omega)$ are incompatible with idempotence of union and intersection $(F \cap F = F; F \cup F = F)$ when there are degrees of membership (Dubois and Prade [5]).

Note further that α-cuts "distribute" over intersection, union, and inclusion. It is easily verified that $\forall \alpha \in]0, 1]$

$$(F \cap G)_\alpha = F_\alpha \cap G_\alpha, \qquad (F \cup G)_\alpha = F_\alpha \cup G_\alpha \qquad (1.51)$$

and that

$$F \subseteq G \Leftrightarrow F_\alpha \subseteq G_\alpha, \qquad \forall \alpha \in]0, 1] \qquad (1.52)$$

Moreover, (1.51) is a further property characterizing min and max for fuzzy intersection and fuzzy union. Equations (1.51) and (1.52) also hold for strong α-cuts.

However, complementation is not interchangeable with α-cut, and we have

$$(\bar{F})_\alpha = \overline{F_{\overline{1 - \alpha}}} \qquad (1.53)$$

[that is, the α-cut of the complement of F is the complement of the strong $(1 - \alpha)$-cut of F].

1.6. Practical Methods for Determining Membership Functions

A question often asked by people beginning study of fuzzy sets is: how are membership functions found? It is appropriate to distinguish the cases where F represents an individual's own idea of a vague category from those where F is determined from statistical data.

1.6.1. Vague Categories as Perceived by an Individual

We must first distinguish between simple categories defined on an objective linear reference scale (e.g., "tall"), and complex categories where several reference scales play a part ("stocky") or where the reference scales may be hard to identify ("handsome").

Let us take simple categories first. In this case, the evaluation of a membership function is a matter of psychometry (see, for example, Krantz *et al.* [19]), and is carried out by questionnaire. Fundamentally, a membership function on Ω defines an ordering "$\geqslant$" on the elements of Ω, and it is the ordering that matters. "$\omega_1 \geqslant \omega_2$" means that "$\omega_1$ is more F than ω_2." Norwich and Turksen [22] have made the connection between classical theories of psychometry and the evaluation of a membership function: if one can define a relation of order in the wide sense, everywhere on Ω, with a largest and a smallest element, then one can represent the category F by a fuzzy set whose membership function, taking values in $\mathbb{R}$, is unique to within a strictly increasing transformation. In order to have a more precise result, a richer ordering structure must be found, namely, an ordering in the wide sense defined on Ω^2. One then compares the pairs (ω_1, ω_1') and (ω_2, ω_2') in the sense of F: $(\omega_1, \omega_1') \geqslant (\omega_2, \omega_2')$ means that "ω_1 is more F compared with ω_1' than ω_2 is F compared with ω_2'." By means of certain axioms of consistency and completeness (particularly that Ω should be a continuum), Norwich and Turksen [22] show that the membership function is unique to within an increasing affine transformation, i.e., unique on $[0, 1]$ once the support and the core of F are known. Practical methods for determining membership functions based on psychometric techniques are being developed (for example, Zimmermann and Zysno [38]).

In practice one can get a rough idea of the form of μ_F which will be adequate for applications. If Ω is the specific reference set for the category, it is easy to elicit from an individual his core $\dot{F}$ and his support $S(F)$. $\dot{F}$ contains all the prototypes of the vague category, while $S(F)$ is obtained by eliminating all objects that do not belong to the category at all. The use of computer graphics can facilitate determination of the behavior of μ_F on $S(F) - \dot{F}$, while avoiding any explicit use of numerical values for the degrees of membership. A further possibility consists of using parametrized forms for μ_F and employing a questionnaire designed to discriminate between parameter values. These two approaches are especially well adapted to determination of fuzzy quantities (see Chapter 2).

Observe that it is not always necessary to obtain exact values for degrees of membership. A slight degree of error in the boundaries of the core or the support, or more generally in the degree of membership, will be of less consequence than when the category has an all-or-nothing representation (an interval), and where the boundaries of the set correspond to points of greatest discontinuity of μ_F. Moreover, the interpretation of these boundaries is not always clear: do they enclose only the prototypes of the category? Do they characterize all objects however slightly involved, or some intermediate set?

A further argument in favor of the idea that an approximate representation of μ_F is adequate in practice is that the error made by combining fuzzy sets by the operations defined above (1.45)–(1.47), and by possibility theory, will never increase since for the most part only max and min operations are used (as in Chapter 2, for calculating fuzzy intervals, for example).

In the more complicated case of a complex category whose reference set is well defined as a Cartesian product of linear scales, the membership function can be obtained by an aggregative process. This is the case, for example, when the category can be described in a branching fashion, by making use of elementary categories and connectives of natural language such as "and," "or," etc. One is thus led to the identification of each simple category, and to the problem of (approximate) determination of fuzzy set operations to represent the connectives. But then one must choose a wider frame than the one defined by (1.45)–(1.47); this is the subject of Chapter 3. See also [38].

Finally, when we have to do with a category whose reference set is difficult to identify (because, being essentially subjective, there is no consensus about it), one can agree on a set made up of a small number of standard values or conditions, possibly ordered, with which to make up a reference set. In such a case it is as well to restrict oneself to a small number of typical degrees of membership.

1.6.2. Fuzzy Sets Constructed from Statistical Data

We can distinguish two aspects. In the first, we have a collection of imprecise data which we wish to model by a frequency distribution for nested families of events. In the second, we wish to approximate a probability distribution in the form of a histogram, by means of a possibility distribution, in such a way that the probabilities of events shall be bracketed between their degrees of possibility and of necessity.

1.6.2.1. Statistics of Imprecise (i.e., Set-Valued) Trials

In drawing up a histogram it is implicitly supposed that distinct data points are mutually exclusive, i.e., given with adequate precision. This hypothesis is not always true: a measuring instrument generally yields intervals, and an opinion poll gives imprecise responses. In such cases, the precise value assumed by a random variable is not accurately ascertained. In other cases, outcomes may fail to be mutually exclusive simply because the possible responses are intrinsically set-valued (e.g., in a survey of foreign languages spoken, a single respondent may speak several languages). Here we suggest how to construct, from such imprecise statistical information, a possibility measure instead of a probability measure, provided the available data are consistent.

We shall suppose that the data are given in the form of closed bounded intervals $\{I_k \mid k = 1, \ldots, q\}$. For imprecise data we shall need a minimum of consistency, namely,

$$I \triangleq \bigcap_{k=1}^{q} I_k \neq \varnothing \tag{1.54}$$

(e.g., the I_k are crisp representatives of the meaning of a vague concept). We can then envisage synthesizing the information in the following way. Take a set of standard nested intervals E_i $(i = 1, \ldots, r)$, such that

$$I \subseteq E_1 \subseteq E_2 \subseteq \cdots \subseteq E_r = \bigcup_{k=1}^{q} I_k$$

The E_i serve as references to classify the data, and have the same function as the disjoint class-intervals used to construct a histogram. Each result I_k is assigned uniquely to the smallest reference E_i capable of including it.

We thus define the frequencies [10]

$$\forall i, \qquad m(E_i) = \frac{1}{q} [\text{number of results assigned to } E_i]$$

It is clear that the function m defines probabilistic weights on the nested focal elements. It therefore defines measures of possibility and necessity [via (1.26) and (1.27), respectively] which bracket the probability values that would have been obtained if the observed results had been precise.

A choice worth considering for the references E_i is the set of α-cuts of the fuzzy set F_* constructed as follows from the set $\{I_k \mid k = 1, q\}$ [39]:

$$\mu_{F_*}(\omega) = \frac{1}{q}[\text{number of } I_k \text{ containing } \omega]$$

Now, with this choice of the E_i, let F^* be the fuzzy set whose membership function is the possibility distribution derived via (1.29) from $\{m(E_i) \mid i = 1, r\}$ as defined above. Then $F_* \subseteq F^*$; moreover, if the I_k are nested, then $F^* = F_*$ is the fuzzy set whose α-cuts are the I_k. See [39] for details: it is shown that F_* and F^* are, in a certain sense, best upper and lower approximations to the data-set $\{I_k \mid k = 1, q\}$. (1.54) ensures that the fuzzy set F_* is normalized.

1.6.2.2. Histograms and Possibility Distributions [9]

Given a possibility measure in the form of nested focal elements with probabilistic weightings, we may seek to approximate to it by means of a probability measure, by interpreting each focal element E_i as a conditional probability $P(\cdot \mid E_i)$ uniformly distributed over E_i. The atom of probability associated with the element $\omega \in \Omega$ (finite) is then

$$\forall \omega, \quad p(\omega) = \sum_{i=1}^{r} P(\omega \mid E_i)m(E_i)$$

$$= \sum_{\omega \in E_i} \frac{m(E_i)}{|E_i|} \tag{1.55}$$

where $|E_i|$ is the number of elements in E_i.

We have therefore made a choice (which could be considered somewhat arbitrary) of one probability measure in the class of all those that satisfy the inequalities [cf. (1.30)]

$$\forall A, \quad N(A) \leqslant P(A) \leqslant \Pi(A) \tag{1.56}$$

The probability atoms $\{p(\omega_i) \mid i = 1, \ldots, n\}$ can be calculated directly from the possibility distribution $\{\pi(\omega_i) \mid i = 1, \ldots, n\}$:

$$p(\omega_i) = \sum_{j=i}^{n} \frac{1}{j}\{\pi(\omega_j) - \pi(\omega_{j+1})\} \tag{1.57}$$

where $\pi(\omega_1) = 1 \geqslant \pi(\omega_2) \geqslant \cdots \geqslant \pi(\omega_{n+1}) = 0$, and ω_{n+1} is a dummy element (Ω has n elements).

It can readily be seen that (1.57) defines a one-to-one correspondence between the distributions p and π. The inverse of this formula is [9]

$$\pi(\omega_i) = \sum_{j=1}^{n} \min(p(\omega_i), p(\omega_j)) \tag{1.58}$$

The latter result allows a fuzzy set to be defined in terms of a histogram, while satisfying the coherence condition (1.56). It is also in this sense that the transformation seems easiest to validate, since one gives up the possibility of summing degrees of uncertainty and is restricted to comparing them. This attitude can only be justified in practice if, for a given problem, a probabilistic model is too difficult to achieve, and if possibility theory can take over its role to yield satisfactory results.

A further feature of this transformation is that it suggests a numerical index of precision for a fuzzy set. By *index of precision* is meant a function from $[0.1]^{\Omega}$ into $[0, 1]$ which is monotone decreasing with respect to fuzzy set inclusion, and which equals 1 only for singleton subsets of the reference set (i.e., the most precise values). An example of index of precision, also called index or measure of *specificity*, has been proposed by Yager [30] for the finite case:

$$\text{Sp}(F) = \int_0^{\bar{\alpha}} \frac{1}{|F_\alpha|} d\alpha \tag{1.59}$$

where $\bar{\alpha} = \sup\{\alpha \mid F_\alpha \neq \varnothing\}$. It is easily seen that

$$\text{Sp}(F) = 1 \qquad \text{if and only if } \exists \, \omega \in \Omega, \qquad F = \{\omega\}$$

and that if F, F' are normalized fuzzy sets,

$$F \subseteq F' \Rightarrow \text{Sp}(F) \geqslant \text{Sp}(F')$$

Putting $\mu_F = \pi$, we establish the relationship between $\text{Sp}(F)$ and the transformation (1.57)–(1.58) because

$$\text{Sp}(F) = \sum_{j=1}^{n} \frac{1}{j} [\pi(\omega_j) - \pi(\omega_{j+1})] = p(\omega_1) \qquad \text{if } \bar{\alpha} = 1$$

where the $\pi(\omega_i)$ are ordered as in (1.57).

We see that $\mathrm{Sp}(F)$ is the probability of the (or of any one, if several) singleton of maximal possibility in the sense of the transformation possibility $\rightarrow$ probability. Further, Higashi and Klir [14] have recently proposed an analog of Shannon's entropy for fuzzy sets, defined by

$$H(F) = \frac{1}{\bar{\alpha}} \int_0^{\bar{\alpha}} \log_2(|F_\alpha|) \, d\alpha$$

$$= \sum_{j=1}^{n} [\pi(\omega_j) - \pi(\omega_{j+1})]\log_2(j) \qquad \text{if } \bar{\alpha} = 1$$

If f is a decreasing one-to-one function from $[0, +\infty)$ into $[0, 1]$, $f \circ H$ has the properties of an index of precision; thus H evaluates imprecision. A simpler way of evaluating the imprecision of a fuzzy set F is to calculate its cardinality $|F|$ defined by De Luca and Termini [2] as

$$|F| = \sum_{i=1}^{n} \mu_F(\omega_i)$$

N.B. An index of precision should not be confused with an index of fuzziness (Kaufmann [16]). The latter evaluates the extent to which the boundaries of a set are ill defined. It is least when F is a sharp subset of Ω and greatest when $\mu_F(\omega) = 0.5$, $\forall \, \omega$. Indices of fuzziness have been introduced in the literature under what seems to us to be the misleading name of "entropy" (De Luca and Termini [2]). $\square$

1.6.3. Remarks on the Set of Degrees of Membership

The evaluation of the membership function is linked with the choice of the set of values for degrees of membership; in the above discussion it was supposed throughout that $\mu_F \in [0, 1]$, which implies that one can always make a linear ordering (in the wide sense) of the elements of Ω in terms of their adaptation to the category F. Some authors have adopted weaker hypotheses: Goguen [11] replaced $[0, 1]$ by a lattice L; Zadeh [34] has suggested that uncertainty about μ_F could be expressed by considering degrees of membership themselves as fuzzy sets on $[0, 1]$. This gives what are called fuzzy sets of type 2. μ_F then has values in the lattice $[0, 1]^{[0, 1]}$; see Mizumoto and Tanaka [20] and Dubois and Prade [4] on this question. Conversely, if it is the element ω, whose degree of membership is $\mu_F(\omega)$, which is imprecise, then Ω is in fact a collection of fuzzy subsets of a different universe V ($\Omega = [0, 1]^V$) and F is a fuzzy set of level 2 (Zadeh [34]) on V. This idea allows vague categories of ever-increasing abstraction to be represented (e.g., "color," regarded as a fuzzy set on {black, gray, red, blue, ...}). See Goguen [12] and Gottwald [13] for systematic studies of this notion.

1.7. Confidence Measures for a Fuzzy Event

A fuzzy (=ill-defined) event can be described by a fuzzy set. We can therefore try to extend the confidence measures defined above to the evaluation of knowledge possessed about the occurrence of a fuzzy event.

If $(\Omega, \mathcal{A}, P)$ is a probability space, where $\mathcal{A}$ is a σ-algebra on Ω, P is a probability measure, and a fuzzy event is described by a measurable membership function μ_F ($\forall \alpha, F_\alpha \in \mathcal{A}$), then the probability of F is defined, according to Zadeh [32], by

$$P(F) = \int_\Omega \mu_F(x) \, dP(x) \tag{1.60}$$

This is the expectation of the membership function.

It can be verified that

$$P(\bar{F}) = 1 - P(F) \tag{1.61}$$

and, also, that

$$P(F \cup G) + P(F \cap G) = P(F) + P(G) \tag{1.62}$$

Further work by Klement [17, 18] makes a systematic study of the concept of fuzzy σ-algebra and establishes under what conditions functions P satisfying (1.62) can be put in the form (1.60).

Along with the probability of a fuzzy event, we can define the notions of possibility and necessity of fuzzy events.

In keeping with the semantics of the concept of possibility, in seeking to measure the possibility of a fuzzy event, one is led, as in Zadeh [37], to utilize an evaluator of the overlap between F and the fuzzy set F_π defined by the possibility distribution π:

$$\Pi(F) = \sup_{\omega \in \Omega} \min(\mu_F(\omega), \pi(\omega)) \tag{1.63}$$

It can be verified that

$$\Pi(F) = 0 \Leftrightarrow S(F) \cap S(F_\pi) = \varnothing$$

$$\Pi(F) = 1 \Leftrightarrow \exists \, \omega = \dot{F} \cap \dot{F}_\pi \quad \text{(finite case)}$$

which generalizes (1.7).

Prade [23] has justified (1.63) in terms of α-cuts, for this definition is equivalent to

$$\Pi(F) = \sup\{\alpha \mid F_\alpha \cap (F_\pi)_\alpha \neq \varnothing\} \qquad (1.64)$$

We can verify that axiom (1.6) for possibilities is still valid for fuzzy events:

$$\Pi(F \cup G) = \max(\Pi(F), \Pi(G)) \qquad (1.65)$$

An immediate consequence of (1.65) is that Π remains a confidence measure on fuzzy events:

$$F \subseteq G \text{ (i.e., } \mu_F \leqslant \mu_G) \Rightarrow \Pi(F) \leqslant \Pi(G)$$

Dually, the necessity of a fuzzy event will be defined by $N(F) = 1 - \Pi(\bar{F})$, or

$$N(F) = \inf_{\omega \in \Omega} \max(\mu_F(\omega), 1 - \pi(\omega)) \qquad (1.66)$$

This notion can be viewed as a degree of inclusion of F_π in F because

$$N(F) = 0 \Leftrightarrow \exists \, \omega \in \overline{S(F)} \cap \dot{F}_\pi \qquad \text{(finite case)}$$
$$N(F) = 1 \Leftrightarrow S(F_\pi) \subseteq \dot{F}$$

We see that the necessity of a fuzzy event involves a notion of inclusion that is stronger than the one defined in (1.43): F contains G whenever the core of F contains the support of G.

It can be verified that if $F_\pi \subseteq F$ in the sense of (1.43), then $N(F) \geqslant 0.5$. Axiom (1.11) for necessity still holds:

$$N(F \cap G) = \min(N(F), N(G)) \qquad (1.67)$$

and the function N is still a confidence measure on fuzzy events.

Let us note that $\max(\Pi(F), \Pi(\bar{F})) \leqslant 1$ whenever F is not a crisp subset of Ω. However, by virtue of (1.49)–(1.50), we always have

$$\max(\Pi(F), \Pi(\bar{F})) \geqslant 0.5, \qquad \min(N(F), N(\bar{F})) \leqslant 0.5 \qquad (1.68)$$

Nevertheless, (1.16) always holds, when F_π is normalized; that is,

$$\forall \, F, \qquad \Pi(F) \geqslant N(F) \qquad (1.69)$$

while (1.17)–(1.18) fail whenever F is not a crisp subset.

1.8. Fuzzy Relations and Cartesian Products of Fuzzy Sets

A *fuzzy relation* is a fuzzy set (or possibility distribution) on a Cartesian product of reference sets. Let Ω_1 and Ω_2 be two reference sets. The fuzzy relation R has membership function $\mu_R = \pi$ whose arguments are $\omega_1 \in \Omega_1$ and $\omega_2 \in \Omega_2$.

The projection of R on Ω_1 is the *marginal possibility distribution* π_1 defined by (Zadeh [35])

$$\pi_1(\omega_1) = \Pi(\{\omega_1\} \times \Omega_2) = \sup_{\omega_2} \pi(\omega_1, \omega_2) \qquad (1.70)$$

This is the analog of a marginal probability distribution, if one regards π as the possibilistic counterpart of a joint probability distribution.

If F_1 is a fuzzy set on Ω_1, then μ_{F_1} can be extended to $\Omega_1 \times \Omega_2$ by

$$\forall\, \omega_1, \omega_2, \qquad \mu_{C_2(F_1)}(\omega_1, \omega_2) = \mu_{F_1}(\omega_1) \qquad (1.71)$$

$C_2(F_1)$ is thus the cylindrical extension of F_1 (Zadeh [35]) to Ω_2. Let F_1 and F_2 be two fuzzy sets on Ω_1 and Ω_2, respectively. The notions of Cartesian product and coproduct can be extended to F_1 and F_2. In the classical case we have the following

Cartesian product:

$$A_1 \times A_2 = \{(\omega_1, \omega_2) \mid \omega_1 \in A_1, \omega_2 \in A_2\}$$

Cartesian coproduct:

$$A_1 + A_2 = \{(\omega_1, \omega_2) \mid \omega_1 \in A_1\} \cup \{(\omega_1, \omega_2) \mid \omega_1 \in A_2\}$$

That is to say,

$$A_1 \times A_2 = C_2(A_1) \cap C_1(A_2)$$
$$A_1 + A_2 = C_2(A_1) \cup C_1(A_2) = \overline{\overline{A_1} \times \overline{A_2}}$$

in terms of cylindrical extensions. Clearly the Cartesian product (or coproduct) of fuzzy sets can be defined as the intersection (or union) of their cylindrical extensions, or (Zadeh [35]) as

$$\mu_{F_1 \times F_2}(\omega_1, \omega_2) = \min(\mu_{F_1}(\omega_1), \mu_{F_2}(\omega_2)) \qquad (1.72)$$

$$\mu_{F_1 + F_2}(\omega_1, \omega_2) = \max(\mu_{F_1}(\omega_1), \mu_{F_2}(\omega_2)) \qquad (1.73)$$

The following remark is important: if $\text{Proj}_i(R)$ is the projection of R onto Ω_i then we always have

$$R \subseteq \text{Proj}_1(R) \times \text{Proj}_2(R) \tag{1.74}$$

where the operation "$\times$" is defined by min, and $\subseteq$ in the sense of (1.43).

Conversely, if F_1 and F_2 are two normalized fuzzy sets on Ω_1 and Ω_2, respectively, then the largest fuzzy relation R such that $F_i = \text{Proj}_i(R)$ is the relation $F_1 \times F_2$, where "$\times$" is defined by min. A fuzzy relation R is called *separable* if and only if

$$R = \text{Proj}_1(R) \times \text{Proj}_2(R) \tag{1.75}$$

N.B. The index H of imprecision of Higashi and Klir introduced above satisfies the remarkable inequality

$$H(R) \leq H(\text{Proj}_1(R)) + H(\text{Proj}_2(R))$$

which is consistent with (1.74); in particular, if R is separable, then equality obtains. □

Variables X_1 and X_2 such that the domain of variation of (X_1, X_2) is delimited by R are then called *noninteractive*. If F_i is the fuzzy set of possible values (in Ω_i) of X_i, X_1 and X_2 are noninteractive if the joint possibility distribution of (X_1, X_2) is

$$\pi_{X_1, X_2} = \min(\mu_{F_1}, \mu_{F_2}) \tag{1.76}$$

Then F_1 and F_2 are called *noninteractive* fuzzy sets. This means that the domain of variation of X_1 is independent of the values assumed by X_2, and conversely. Equation (1.76) defines the largest (in the sense of inclusion) fuzzy relation, or equivalently the least constrained possibility function, which can have F_1 and F_2 as projections. In other words, if F_1 and F_2 are the only constraints acting on (X_1, X_2), (1.76) is the least arbitrary way to define π_{X_1, X_2}. Equation (1.76) no longer holds as soon as X_1 and X_2 are linked by any relation whatever [e.g., if $\exists f: X_2 = f(X_1)$].

Let Π_{12} be a possibility measure defined in terms of a distribution π_{12} on $\Omega_1 \times \Omega_2$. If π_{12} is separable and normalized, then

$$\forall A_1 \times A_2 \subseteq \Omega_1 \times \Omega_2, \quad \Pi_{12}(A_1 \times A_2) = \min(\Pi_1(A_1), \Pi_2(A_2)) \tag{1.77}$$

where Π_i is defined from the projection of π_{12} on Ω_i, and

$$N_{12}(A_1 + A_2) = \max(N_1(A_1), N_2(A_2)) \qquad (1.78)$$

where N_{12}, N_1, and N_2 are the necessity measures dual to Π_{12}, Π_1, and Π_2, respectively. In any case, if π_{12} is separable and normalized, (1.77) and (1.78) can be extended to fuzzy events F_1 and F_2 in Ω_1 and Ω_2, respectively, that is (Prade [23]),

$$\forall (F_1, F_2) \in [0, 1]^{\Omega_1} \times [0, 1]^{\Omega_2}$$

$$\Pi_{12}(F_1 \times F_2) = \min(\Pi_1(F_1), \Pi_2(F_2)) \qquad (1.79)$$

$$N_{12}(F_1 + F_2) = \max(N_1(F_1), N_2(F_2)) \qquad (1.80)$$

Observe that even if π_{12} is not separable, and whether F_1, F_2 are fuzzy or not, we of course have

$$\Pi_{12}(F_1 + F_2) = \max(\Pi_1(F_1), \Pi_2(F_2)) \qquad (1.81)$$

$$N_{12}(F_1 \times F_2) = \min(N_1(F_1), N_2(F_2)) \qquad (1.82)$$

while (1.79) and (1.80) become inequalities ($\leq$ and $\geq$, respectively).

It will be seen that there is a certain analogy between stochastic independence and "possibilistic" noninteraction. If P_{12} is a joint probability measure on $\Omega_1 \times \Omega_2$, P_1 and P_2 are the marginal probability measures for the variables X_1 and X_2, respectively, then if the two variables X_1 and X_2 are stochastically independent we have

$$\forall A_1 \subseteq \Omega_1, \qquad \forall A_2 \subseteq \Omega_2, \qquad P_{12}(A_1 \times A_2) = P_1(A_1)P_2(A_2) \quad (1.83)$$

which expresses, on the frequentist interpretation, that A_1 occurs with the same frequency among the occasions when A_2 also occurs as among the occasions when A_2 does not occur (and conversely). On the other hand, (1.76) means that the domain of variation of X_1 is independent of that of X_2 and conversely. However, if for two events A_1 and A_2 we have the relation

$$P_{12}(A_1 \times A_2) = \min(P_1(A_1), P_2(A_2)) \qquad (1.84)$$

then we can deduce that A_1 implies A_2 [if $P_1(A_1) \leq P_2(A_2)$] or, on the frequentist interpretation, that A_1 occurs only on occasions when A_2 occurs. In fact, since $P_1(A_1) = P_{12}(A_1 \times A_2) + P_{12}(A_1 \times \bar{A}_2)$ and $P_2(A_2) = P_{12}(A_1 \times A_2) + P_{12}(\bar{A}_1 \times A_2)$, (1.84) implies that either

$P_{12}(A_1 \times \bar{A}_2) = 0$ or $P_{12}(\bar{A}_1 \times A_2) = 0$. Therefore (1.84) means that A_1 and A_2 are anything but independent!

References

1. DEMPSTER, A. P. (1967). Upper and lower probabilities induced by a multivalued mapping. *Ann. Math. Stat.*, **38**, 325–339.
2. DE LUCA, A., and TERMINI, S. (1972). A definition of a non-probabilistic entropy, in the setting of fuzzy sets theory. *Inf. Control*, **20**, 301–312.
3. DUBOIS, D. (1983). Modèles Mathématiques de l'Imprécis et de l'Incertain en Vue d'Applications aux Techniques d'Aide à la Décision. Thesis, Scientific and Medical University of Grenoble.
4. DUBOIS, D., and PRADE, H. (1979). Operations in a fuzzy-valued logic. *Inf. Control*, **43**(2), 224–240.
5. DUBOIS, D., and PRADE, H. (1980). New results about properties and semantics of fuzzy set-theoretic operators. In *Fuzzy Sets: Theory and Applications to Policy Analysis and Information Systems* (Paul P. Wang and S. K. Chang, eds.). Plenum, New York, pp. 59–75.
6. DUBOIS, D., and PRADE, H. (1980). *Fuzzy Sets and Systems: Theory and Applications*, Vol. 144 in Mathematics in Sciences and Engineering Series. Academic, New York.
7. DUBOIS, D., and PRADE, H. (1982). Towards fuzzy differential calculus. Part 2: integration on fuzzy intervals. *Fuzzy Sets Syst.* **8**, 105–116.
8. DUBOIS, D., and PRADE, H. (1982). On several representations of an uncertain body of evidence. In *Fuzzy Information and Decision Processes* (M. M. Gupta and E. Sanchez, eds.), North-Holland, Amsterdam, pp. 167–181.
9. DUBOIS, D., and PRADE, H. (1983). Unfair coins and necessity measures: Towards a possibilistic interpretation of histograms. *Fuzzy Sets Syst.*, **10**(1), 15–20.
10. DUBOIS, D., and PRADE, H., (1984). The statistical approach to membership functions: random experiments with imprecise outcomes. In *Proceedings of the Workshop on the Membership Function*, Brussels, March 22–23, 1984, EIASM, Brussels, pp. 51–61. Extended version DUBOIS, D., and PRADE, H. (1986). Fuzzy sets and statistical data. *Eur. J. Oper. Res.* **25**, 345–356.
11. GOGUEN, J. A. (1967). L-Fuzzy sets. *J. Math. Anal. Appl.*, **18**(1), 145–174.
12. GOGUEN, J. A. (1974). Concept representation in natural and artificial languages: Axioms, extensions and applications for fuzzy sets. *Int. J. Man-Machine Stud.*, **6**, 513–561.
13. GOTTWALD, S. (1979). Set theory for fuzzy sets of higher levels. *Fuzzy Sets and Syst.*, **2**, 125–151.
14. HIGASHI, M., and KLIR, G. J. (1983). Measures of uncertainty and information based on possibility distributions. *Int. J. Gen. Syst.*, **9**(1), 43–58.
15. HUBER, P. J. (1973). The use of Choquet capacities in statistics. *Bull. Int. Stat. Inst.*, **XLV**(4), 181–188.
16. KAUFMANN, A. (1973). *Introduction à la Théorie des Sous-Ensembles Flous*. Vol. 1: *Eléments Théoriques de Base*. Masson, Paris.
17. KLEMENT, E. P. (1980). Characterisation of finite fuzzy measures using Markoff-kernels. *J. Math. Anal. Appl.*, **75**, 330–339.
18. KLEMENT, E. P. (1982). Construction of fuzzy σ-algebras using triangular norms, *J. Math. Anal. Appl.*, **85**, 543–566.

19. KRANTZ, D. H., LUCE, R., SUPPES, P., and TVERSKI, A. (1971). *Foundations of Measurement.* Vol. 1, Academic, New York.
20. MIZUMOTO, M., and TANAKA, K. (1976). Some properties of fuzzy sets of type 2. *Inf. Control,* **31,** 312–340.
21. NGUYEN, H. T. (1979). Some mathematical tools for linguistic probabilities. *Fuzzy Sets Syst.,* **2,** 53–65.
22. NORWICH, A. M., and TURKSEN, I. B. (1982). The fundamental measurement of fuzziness. In *Fuzzy Set and Possibility Theory: Recent Developments* (R. R. Yager, ed.). Pergamon, New York, pp. 49–60.
23. PRADE, H. (1982). Modal semantics and fuzzy set theory. In *Fuzzy Set and Possibility Theory: Recent Developments* (R. R. Yager, ed.). Pergamon, New York, pp. 232–246.
24. PRADE, H. (1982). Modèles Mathématiques de l'Imprécis et de l'Incertain en Vue d'Applications au Raisonnement Naturel. Thesis No. 1048, Paul Sabatier University, Toulouse.
25. PURI, M. L., and RALESCU, D. (1982). A possibility measure is not a fuzzy measure. *Fuzzy Sets Syst.,* **7,** 311–314.
26. RALESCU, D. (1979). A survey of the representation of fuzzy concepts and its applications. In *Advances in Fuzzy Set Theory and Applications* (M. M. Gupta, R. K. Ragade, and R. R. Yager, eds.). North-Holland, Amsterdam, pp. 77–91.
27. SAVAGE, L. J. (1972). *The Foundations of Statistics,* Dover, New York.
28. SHAFER, G. (1976). *A Mathematical Theory of Evidence.* Princeton University Press, Princeton, New Jersey.
29. SUGENO, M. (1974). Theory of Fuzzy Integrals and Its Applications. Ph.D. thesis, Tokyo Institute of Technology, Japan.
30. YAGER, R. R. (1982). Measuring tranquility and anxiety in decision-making: An application of fuzzy sets. *Int. J. Man-Machine Stud.,* **8,** (3), 139–146.
31. ZADEH, L. A. (1965). Fuzzy sets. *Inf. Control,* **8,** 338–353.
32. ZADEH, L. A. (1968). Probability measures of fuzzy events. *J. Math. Anal. Appl.,* **23,** 421–427.
33. ZADEH, L. A. (1971). Similarity relations and fuzzy orderings. *Inf. Sci.,* **3,** 177–200.
34. ZADEH, L. A. (1971). Quantitative fuzzy semantics. *Inf. Sci.,* **3,** 159–176.
35. ZADEH, L. A. (1975). Calculus of fuzzy restrictions. In *Fuzzy Sets and Their Applications to Cognitive and Decision Processes* (L. A. Zadeh, K. S. Fu, K. Tanaka, and M. Shimura, eds.). Academic, New York, pp. 1–39.
36. ZADEH, L. A. (1975). The concept of a linguistic variable and its applications to approximate reasoning. *Inf. Sci.* Part 1, **8,** 199–249; Part 2, **8,** 301–357; Part 3, **9,** 43–80.
37. ZADEH, L. A. (1978). Fuzzy sets as a basis for a theory of possibility. *Fuzzy Sets Syst.,* **1,** 3–28.
38. ZIMMERMANN, H. J., and ZYSNO, P. (1983). Decisions and evaluations by hierarchical aggregation of information. *Fuzzy Sets Syst.,* **10**(3), 243–260.
39. DUBOIS, D., and PRADE, H. (1986). A set-theoretic view of belief functions: Logical operations and approximations by fuzzy sets. *Int. J. Gen. Syst.* **12,** 193–226.

2

The Calculus of Fuzzy Quantities

This chapter gives methods of calculation for expressions containing imprecise quantities, represented by possibility distributions on the real numbers. These methods are in complete agreement with what is commonly called interval analysis, of which they constitute an extension to the case of weighted intervals. Their usefulness is illustrated by some examples at the end of the chapter. Moreover, fuzzy quantities will enter extensively in Chapters 3, 5, and 6. In essence, the calculus of fuzzy quantities constitutes a refinement of sensitivity analysis, which thereby acquires nuance, and this *without great increase* in the amount of calculation required. The calculus of fuzzy quantities can replace the calculus of random functions (cf. Papoulis [21]) when this proves too intractable, though of course with more or less loss of information according to the type of problem. A more detailed account of the theoretical part of this chapter may be found in Ref. 27. An introductory text is Ref. 28.

2.1. Definitions and a Fundamental Principle

2.1.1. Fuzzy Quantities, Fuzzy Intervals, Fuzzy Numbers

A *fuzzy quantity* Q is a fuzzy set on the reals, that is to say a mapping μ_Q from $\mathbb{R}$ into $[0, 1]$. μ_Q will naturally be viewed as a possibility distribution on the values that a variable can assume. In accordance with the discussion in Section 1.4, concerning the normalization of fuzzy sets and the (by definition) "possibilistic" nature of the fuzzy quantity Q, we shall suppose, unless otherwise stated, that μ_Q is normalized.

Any real number m in the core $\dot{Q}$ [i.e., such that $\mu_Q(m) = 1$] is called a *modal value* of Q.

We can define a kind of fuzzy quantity that generalizes the notion of interval: a *fuzzy interval* is a convex fuzzy quantity, that is, one whose membership function is quasiconcave (Zadeh [23]):

$$\forall u, v, \qquad \forall w \in [u, v], \qquad \mu_Q(w) \geqslant \min(\mu_Q(u), \mu_Q(v)) \qquad (2.1)$$

A fuzzy quantity is convex if and only if its α-cuts are convex, i.e., they are intervals (bounded or not). Closed intervals are generalized by fuzzy intervals whose membership functions are upper semicontinuous (u.s.c), i.e., by definition, whose α-cuts are closed intervals. The compact (closed and bounded) subsets of $\mathbb{R}$ are generalized by u.s.c. fuzzy quantities with compact support. An u.s.c. interval with compact support and unique modal value will be called a *fuzzy number*. If M is a fuzzy number with modal value m, M is a possible representation of "about m" (Dubois and Prade [6]). In the case of a fuzzy interval, the set of modal values is an interval. A fuzzy quantity Q will be called *multimodal* if there is a *finite* collection of convex u.s.c. fuzzy sets $\{M_i \mid i \in I\}$ such that Q is the union of the M_i in the sense of (1.47).

A fuzzy interval is a convenient representation of an imprecise quantity, richer in information than a precise interval. In practice one must often estimate the precision of a parameter, or express one's desire concerning what value an attribute should have, but where an interval is unsatisfactory. Must one then set the boundaries of the interval pessimistically (in which case it will be wide, and subsequent calculations will be uninformative) or optimistically (in which case the value thus delimited may fall outside the domain assigned and the apparent precision of the results be illusory)? A fuzzy interval allows a representation that is at once both pessimistic and optimistic: let the support be chosen so as to be sure that the delimited value will not fall outside it, and the core so as to include the most plausible values.

The procedures for determining membership functions, from the imprecise perceptions of an individual or from statistical data, which were discussed in Section 1.6, are especially applicable to fuzzy quantities.

In terms of the probabilistic interpretation of membership functions (see Section 1.3.2), the mean value of a fuzzy interval can be rigorously defined, and it turns out to be an interval [11, 26].

Let Π be the possibility measure associated with the distribution μ_Q, where Q is an u.s.c. fuzzy quantity with compact support. Note that in this case the possibility measure satisfies the continuity condition (1.3) for confidence measures on monotone increasing or decreasing sequences of

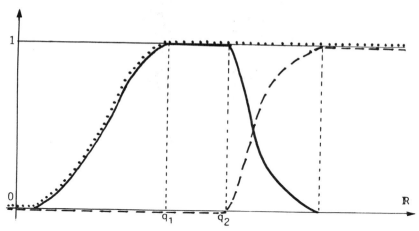

Figure 2.1. Upper and lower distribution functions. ——, Q; $\cdots$, F^*; ---, F_*.

compact sets. Consider the set $\mathscr{P}$ of probability measures compatible with Π, that is [cf. equation (1.30)],

$$\mathscr{P} = \{P \mid \forall A, N(A) \leq P(A) \leq \Pi(A)\}$$

Let $[q_1, q_2]$ be the core $\dot{Q}$ of Q.

The upper bound of $\mathscr{P}$ is attained by the probability measure $\bar{P}$ whose distribution function F^* is such that (cf. Figure 2.1)

$$\forall u \in \mathbb{R}, \qquad F^*(u) = \bar{P}((-\infty, u]) = \Pi((-\infty, u]) = \sup\{\mu_Q(r) \mid r \leq u\},$$

i.e.,

$$F^*(u) = \mu_Q(u) \qquad \text{if } u \leq q_1$$

$$= 1 \qquad \text{if } u > q_1$$

Likewise, the lower bound of $\mathscr{P}$ is attained by the probability measure $\underline{P}$, whose distribution function is such that (cf. Figure 2.1)

$$\forall u \in \mathbb{R}, \qquad F_*(u) = \underline{P}((-\infty, u]) = N((-\infty, u])$$

$$= \inf\{1 - \mu_Q(r) \mid r > u\},$$

i.e.,

$$F_*(u) = 0 \qquad \text{if } u < q_2$$

$$= 1 - \lim_{r \searrow u} \mu_Q(r) \qquad \text{if } u \geq q_2$$

Applying the definitions of upper and lower expectations given in (1.31) and (1.32), the lower and upper mean values of the fuzzy quantity Q can be defined, respectively, as

$$E_*(Q) = \int_{-\infty}^{+\infty} r \, d\Pi((-\infty, r]) = \int_{-\infty}^{+\infty} r \, dF^*(r) \qquad (2.2)$$

$$E^*(Q) = \int_{-\infty}^{+\infty} r \, dN((-\infty, r]) = \int_{-\infty}^{+\infty} r \, dF_*(r) \qquad (2.3)$$

The mean value of a fuzzy interval Q will therefore be the collection of mean values of all the random variables compatible with Q [i.e., which satisfy (1.30)], namely, the interval $[E_*(Q), E^*(Q)]$: it seems natural that the mean of a fuzzy interval should be an interval. If $Q = [a, b]$ then it is easily verified that $E_*(Q) = a$, $E^*(Q) = b$.

2.1.2. The Extension Principle

In this section we pose the following question: given several noninteracting (cf. Section 1.8) possibilistic variables $X, Y, Z, \ldots$, each defined by a fuzzy quantity that restricts its domain, how may one calculate the fuzzy quantity that restricts the domain of the variable $f(X, Y, Z, \ldots)$, where f is a given real function?

For this, consider two sets Ω and U, and a mapping from Ω into U. Suppose there is a confidence measure g on $(\Omega, \mathcal{A})$, where $\mathcal{A}$ is the collection of subsets of Ω on which g is defined. Then, from the inverse image of a subset A of U, defined by $f^{-1}(A) = \{\omega \mid f(\omega) \in A\}$, we can construct a set function g_f, obtained from g through f, which is given by

$$\forall A \in \mathcal{S}, \qquad g_f(A) = g(f^{-1}(A)) \qquad (2.4)$$

where $\mathcal{S} = \{A \mid f^{-1}(A) \in \mathcal{A}\}$. It is easily verified that g_f is a confidence measure on $(U, \mathcal{S})$.

A particular and well-known case of (2.4) is that where g is a probability measure. If $\Omega = U = \mathbb{R}$, and g is a probability measure on $\mathbb{R}$ (furnished with the Borel σ-algebra), defining a random variable X, then f is a function of random variables. Knowing the distribution function F_X associated with g [$F_X(r) = \text{Prob}(X \leq r)$], we are interested in the distribution function F_Y where $Y = f(X)$ defined by $F_Y(r) = \text{Prob}[f(X) \leq r]$ (see, for example, Papoulis [21]). In this book, we consider the case where g is a possibility measure.

Equation (2.4) becomes

$$\Pi_F(A) = \Pi(f^{-1}(A))$$
$$= \sup\{\pi(\omega) \,|\, f(\omega) \in A\} \qquad \text{if } f^{-1}(A) \neq \emptyset \qquad (2.5)$$

where π is the possibility distribution associated with Π. Note that we may choose $\mathscr{A} = 2^{\Omega}$ (so $\mathscr{S} = 2^U$).

Π_f is certainly a possibility measure since $f^{-1}(A \cup B) = f^{-1}(A) \cup f^{-1}(B)$. The possibility distribution π_f associated with Π_f is given by

$$\forall u \in U, \qquad \pi_f(u) = \Pi_f(\{u\}) = \sup\{\pi(\omega) \,|\, f(\omega) = u\}$$
$$= 0 \qquad \text{if } f^{-1}(u) = \emptyset \qquad (2.6)$$

Equation (2.6) is known in fuzzy set theory as the *extension principle* (Zadeh [24, 25]).

We can apply (2.4) to the necessity measure N dual to Π. The set function N_f thus constructed is certainly a necessity measure, and is the dual measure of Π_f since $f^{-1}(\bar{A}) = \overline{f^{-1}(A)}$ when the mapping f is onto. N_f can also be expressed in terms of π_f through (1.14).

When Ω is a Cartesian product $\Omega_1 \times \Omega_2$ and π is separable, i.e., expressible as $\pi = \min(\mu_{Q_1}, \mu_{Q_2})$, where Q_1 and Q_2 are fuzzy sets restricting the domains of variation of the noninteractive variables X_1 and X_2 (cf. Section 1.8), then the extension principle can be written

$$\pi_f(u) = \sup\{\min(\mu_{Q_1}(\omega_1), \mu_{Q_2}(\omega_2)) \,|\, f(\omega_1, \omega_2) = u\}$$
$$= 0 \qquad \text{if } f^{-1}(u) = \emptyset \qquad (2.7)$$

π_f is the possibility distribution restricting the domain of the variable $f(X_1, X_2)$, namely a fuzzy set denoted by $f(Q_1, Q_2)$, with membership function $\mu_{f(Q_1, Q_2)} = \pi_f$. It is clear that when Q_1 and Q_2 reduce to singletons $\{\omega_1\}$ and $\{\omega_2\}$, $f(Q_1, Q_2)$ is the singleton $\{f(\omega_1, \omega_2)\}$.

The fuzzy set $f(Q_1, Q_2)$ can be constructed from α-cuts $(Q_1)_\alpha$ and $(Q_2)_\alpha$, since (Nguyen [20])

$$\forall u, \qquad \mu_{f(Q_1, Q_2)}(u) = \sup\{\alpha \in \,]0, 1] \,|\, u \in f((Q_1)_\alpha, (Q_2)_\alpha)\} \quad (2.8)$$

However, we do not in general have

$$(f(Q_1, Q_2))_\alpha = f((Q_1)_\alpha, (Q_2)_\alpha) \qquad \text{for all } \alpha \in \,]0, 1]$$

We only have

$$f((Q_1)_\alpha, (Q_2)_\alpha) \subseteq [f(Q_1, Q_2)]_\alpha \qquad (2.9)$$

Nguyen [20] has shown that equality in (2.9) obtains if and only if the supremum is attained in (2.7) for all $u \in U$.

Moreover, if we consider strong α-cuts, we always have the equality (Negoita [19]):

$$f((Q_1)_{\tilde{\alpha}}, (Q_2)_{\tilde{\alpha}}) = [f(Q_1, Q_2)]_{\tilde{\alpha}} \qquad (2.10)$$

The extension principle (2.6) can naturally be generalized to the case where f is a fuzzy relation R, viewed as a multivalued function that maps ω into a fuzzy set $f(\omega) \subseteq U$ of possible images, defined by

$$\forall u, \qquad \mu_{f(\omega)}(u) = \mu_R(\omega, u) \qquad (2.11)$$

This generalization can be written

$$\mu_{Q \circ R}(u) = \sup\{\min(\mu_Q(\omega), \mu_R(\omega, u)) \mid \omega \in \Omega\} \qquad (2.12)$$

and is known as the "sup–min composition," denoted by "$\circ$", the notations $Q \circ R$ and $f(Q)$ being equivalent, by (2.11).

Let S be a fuzzy relation on $U \times V$. We then have the associative property

$$(Q \circ R) \circ S = Q \circ (R \circ S)$$

where $R \circ S$ is the fuzzy relation relation on $\Omega \times V$ defined by

$$\mu_{R \circ S}(\omega, v) = \sup\{\min(\mu_R(\omega, u), \mu_S(u, v)) \mid u \in U\}$$

A remarkable special case of this associativity is obtained on taking for R and S univalent functions f and g, when we have, symbolically, $(g \circ f)(Q) = g(f(Q))$. See Ref. 29 for other fuzzy relational products.

2.2. Calculus of Fuzzy Quantities with Noninteractive Variables

In this section, we give practical sufficient conditions for the α-cut of $f(Q_1, Q_2)$ to be expressible in terms of the α-cuts of Q_1 and Q_2, where $f(Q_1, Q_2)$ is a fuzzy quantity obtained by applying the extension principle to a real function f of real variables, and Q_1, Q_2 are two noninteracting fuzzy quantities. This result will demonstrate how the calculus of fuzzy intervals generalizes the calculus of errors and thus also the calculus of real numbers. In addition it lies at the base of the practical methods of calculation which are the subject of the following section.

2.2.1. Fundamental Result

Proposition 2.1 (Dubois [4]). Let M and N be two fuzzy intervals with u.s.c. membership functions. We suppose that the α-cuts M_α and N_α, for all $\alpha > 0$, do not include the entire set $\mathbb{R}$ of reals. Let f be a continuous isotonic mapping from $\mathbb{R}^2$ into $\mathbb{R}$, that is

$$\forall u \geqslant u', \qquad \forall v \geqslant v', \qquad f(u, v) \geqslant f(u', v')$$

Then *the α-cuts of the fuzzy quantity $f(M, N)$ are the images under f of the α-cuts of M and N.* Mathematically,

$$\forall \alpha > 0, \qquad [f(M, N)]_\alpha = f(M_\alpha, N_\alpha) \tag{2.13}$$

If M_α and N_α are closed bounded intervals of the form $[\underline{m}_\alpha, \bar{m}_\alpha]$ and $[\underline{n}_\alpha, \bar{n}_\alpha]$, then

$$\forall \alpha \in \,]0, 1], \qquad f(M_\alpha, N_\alpha) = [f(\underline{m}_\alpha, \underline{n}_\alpha), f(\bar{m}_\alpha, \bar{n}_\alpha)]$$

i.e., is a closed interval.

N.B. Equation (2.13) is proved by Nguyen[20] for the case of fuzzy quantities with compact support and u.s.c. membership function, for any continuous function f; Nguyen's results still hold if all the α-cuts of the fuzzy quantities involved are compact. □

Equation (2.13) also holds under the following conditions (in addition to the continuity of f):

- f is defined only on a rectangle [pavé] of $\mathbb{R}$ (in which case we consider the restriction of μ_M and μ_N to this rectangle).
- f is antitonic: $u \geqslant u'$ and $v \geqslant v' \Rightarrow f(u, v) \leqslant f(u', v')$, in which case, $f(M_\alpha, N_\alpha) = [f(\bar{m}_\alpha, \bar{n}_\alpha, f(\underline{m}_\alpha, \underline{n}_\alpha)]$ when M_α and N_α are closed and bounded.
- f is "hybrid": $u \geqslant u'$ and $v \leqslant v' \Rightarrow f(u, v) \geqslant f(u', v')$, in which

case $f(M_\alpha, N_\alpha) = [f(\underline{m}_\alpha, \bar{n}_\alpha), f(\bar{m}_\alpha, \underline{n}_\alpha)]$ when M_α and N_α are closed and bounded.

• f has more than two arguments and is monotone in each of them.

Thus it can be seen that when M_α and N_α are closed and bounded for all α, $(f(M, N))_\alpha$ is closed and bounded under the conditions of Proposition 2.1; therefore $f(M, N)$ has an u.s.c. membership function. Moreover, since M and N are fuzzy intervals, it is clear that $f(M, N)$ is also a fuzzy interval, still under the conditions of Proposition 2.1.

A remarkable consequence of Proposition 2.1 is that the calculation of $f(M, N)$ can be done by combining two parts separately calculated: on the one hand, the increasing parts of μ_M and μ_N (μ_M^+ and μ_N^+, say, defined on $[-\infty, \underline{m}_1]$ and on $[-\infty, \underline{n}_1]$, respectively); and on the other hand the decreasing parts (μ_M^- and μ_N^-, defined on $[\bar{m}_1, +\infty)$ and $[\bar{n}_1, +\infty)$), in the case of isotonic or antitonic functions, where the intervals $[\underline{m}_1, \bar{m}_1]$ and $[\underline{n}_1, \bar{n}_1]$ are the sets of modal values of M and N.

The following table summarizes the four possible cases, for a function f of two arguments:

$f(x, y)$		$\mu_{f(M,N)}^+$ made up	
in x	in y	with	and
Increasing	Increasing	μ_M^+	μ_N^+
Increasing	Decreasing	μ_M^+	μ_N^-
Decreasing	Increasing	μ_M^-	μ_N^+
Decreasing	Decreasing	μ_M^-	μ_N^-

Note: $\mu_{f(M,N)}^-$ is obtained by interchanging $+$ and $-$ in the above table.

This method of calculation is illustrated in Figure 2.2 for the case of an isotonic function.

Thus, when μ_M^+, μ_M^-, μ_N^+, and μ_N^- are strictly monotonic, we can write

$$\forall\, w \leq f(\underline{m}_1, \underline{n}_1),$$
$$\mu_{f(M,N)}^+(w) = \sup\{\min\{\mu_M^+(\underline{m}_\alpha), \mu_N^+(\underline{n}_\alpha)\}, \alpha \in]0, 1], f(\underline{m}_\alpha, \underline{n}_\alpha) = w\} \tag{2.14}$$

$$\forall\, w \in [f(\underline{m}_1, \underline{n}_1), f(\bar{m}_1, \bar{n}_1)], \qquad \mu_{f(M,N)}(w) = 1 \tag{2.15}$$

$$\forall\, w \geq f(\bar{m}_1, \bar{n}_1),$$
$$\mu_{f(M,N)}^-(w) = \sup\{\min\{\mu_M^-(\bar{m}_\alpha), \mu_N^-(\bar{n}_\alpha)\}, \alpha \in]0, 1], f(\bar{m}_\alpha, \bar{n}_\alpha) = w\} \tag{2.16}$$

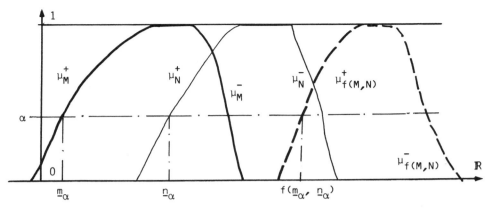

Figure 2.2. Performing an operation on fuzzy intervals.

When f is strictly isotonic $[u > u', v > v' \Rightarrow f(u, v) > f(u', v')]$, then if μ_M^+, μ_N^+ are strictly increasing, and μ_M^-, μ_N^- are strictly decreasing, (2.14) and (2.16) give (Dubois and Prade [7])

$$\forall \varepsilon \in \{-, +\}, \qquad (\mu_{f(M,N)}^\varepsilon)^{-1} = f((\mu_M^\varepsilon)^{-1}, (\mu_N^\varepsilon)^{-1}) \qquad (2.17)$$

N.B. Equation (2.17) remains valid when μ_M or μ_N has plateaus of constant value outside the core, if the notion of inverse function is extended, these plateaus corresponding to discontinuities of $(\mu_M^\varepsilon)^{-1}$ or $(\mu_N^\varepsilon)^{-1}$.   □

The relation (2.17) clearly shows that applying the extension principle to a large class of numerical functions leads to simple explicit calculations. To obtain $f(M, N)$ it suffices to solve for α the equation

$$w = f((\mu_M^\varepsilon)^{-1}(\alpha), (\mu_N^\varepsilon)^{-1}(\alpha)), \qquad \varepsilon \in \{+, -\} \qquad (2.18)$$

Nevertheless, if (2.18) should be difficult to solve, one can make do with

- either using the inverse functions of $\mu_{f(M,N)}^+$ and $\mu_{f(M,N)}^-$ obtained from (2.17) in *analytic form*;
- or discretizing $[0, 1]$ and calculating $f(M_\alpha, N_\alpha)$ for a finite number of α-values, thus obtaining a sample of values of the membership function of $f(M, N)$.

In the case where Q_1 and Q_2 are multimodal fuzzy quantities, the evaluation of $f(Q_1, Q_2)$ can be reduced to calculations on fuzzy intervals by means of the following two simple results (Dubois and Prade [7]):

Proposition 2.2. If Q_1 and Q_2 are two fuzzy sets in $\mathbb{R}$ such that $\sup \mu_{Q_1} \leqslant \sup \mu_{Q_2}$, then if Q_2 is the truncation of Q_2 at level $\sup \mu_{Q_1}$ [i.e., $\mu_{Q_2} = \min(\sup \mu_{Q_1}, \mu_{Q_2})$], we have

$$\forall f: \qquad f(Q_1, Q_2) = f(Q_1, Q_2) \qquad \text{(truncation effect)}$$

Proposition 2.3. If

$$Q_1 = \bigcup_{i=1}^{m_1} M_i^1, \qquad Q_2 = \bigcup_{j=1}^{m_2} M_j^2$$

where M_i^1 and M_j^2 are convex fuzzy sets in $\mathbb{R}$, then

$$f(Q_1, Q_2) = \bigcup_{i=1}^{m_1} \bigcup_{j=1}^{m_2} f(M_i^1, M_j^2)$$

A multimodal fuzzy quantity is a finite union of fuzzy intervals (possibly nonnormalized). These two results allow the calculation of $f(Q_1, Q_2)$ as a union of fuzzy intervals obtained by combining each of the intervals making up Q_1 with the intervals making up Q_2. This combination is done with fuzzy intervals which are of the same height, owing to the truncation effect.

2.2.2. Relation to Interval Analysis

If two quantities M and N are known imprecisely, in the form of ordinary subsets of $\mathbb{R}$, the quantity $f(Q_1, Q_2)$ is defined by

$$f(Q_1, Q_2) = \{f(u, v) \mid u \in Q_1, v \in Q_2\} \qquad (2.19)$$

When Q_1 and Q_2 are closed intervals we have the interval analysis, which is well known in physics. Interval analysis has experienced a revival of interest, with the arrival of computers, in connection with the evaluation of rounding errors. The books by Moore [16, 17] give a survey of the subject. It is easily seen that (2.19) is equivalent to the extension principle (2.7) when Q_1 and Q_2 are sets of reals. The formula (2.8) shows that the sup–min extension principle amounts to defining $f(Q_1, Q_2)$, according to the representation formula (1.36), as a family of nested sets $f(Q_{1\alpha}, Q_{2\alpha})$ obtained, for intervals, by the methods of interval analysis. By (2.10), one can equally use the family of sets $f(Q_{1\bar{\alpha}}, Q_{2\bar{\alpha}})$ to represent

$f(Q_1, Q_2)$. In this case, a fuzzy interval Q is to be viewed as a family $J(Q)$ of confidence intervals (the strong α-cuts of Q). The value α is the degree of possibility for the variable whose value is constrained by Q to lie outside $Q_{\bar{\alpha}}$, i.e., the complement with respect to 1 of the degree of necessity that the variable constrained by Q should lie inside. In fact, if $S \in J(Q)$, then $S = Q_{\bar{\alpha}}$, where α is defined by

$$1 - \alpha = N(S) = \inf\{1 - \mu_Q(\mu) \mid \mu \notin S\} \qquad (2.20)$$

The book by Moore [17] will be used as reference, in the following, to show that all the structural properties of the calculus of closed intervals also hold for u.c.s. fuzzy intervals. For example, the monotonicity property

$$Q_1 \subseteq Q_1', \qquad Q_2 \subseteq Q_2' \Rightarrow f(Q_1, Q_2) \subseteq f(Q_1', Q_2') \qquad (2.21)$$

noted by Moore [17] for intervals, carries over to fuzzy intervals. This result means that as the imprecision of the data increases, the imprecision of the results derived from them can only increase also.

Because it agrees with interval analysis, the calculus of fuzzy quantities can be called "pessimistic" since $f(Q_1, Q_2)$ is [in the inclusion sense of (1.43)] the *maximal* fuzzy set such that the arguments of f are constrained by Q_1 and Q_2, respectively. This is easily verified by noting that each α-cut of $f(Q_1, Q_2)$ is the greatest possible, in accordance with (2.19).

2.2.3. Application to Standard Operations

The general results presented above allow the demonstration of some interesting properties when they are applied to standard operations, especially the four arithmetic operations, and also max and min. We first consider unary operations.

2.2.3.1. Functions of a Real Variable

If f has just one argument and Q is a fuzzy quantity, $f(Q)$ has the membership function

$$\mu_{f(Q)}(w) = \sup_{w=f(u)} \mu_Q(u)$$

$$= 0 \qquad \text{if } f^{-1}(w) = \varnothing$$

$$= \mu_Q(f^{-1}(w)) \qquad \text{if } f \text{ is "into" (injective)} \qquad (2.22)$$

Thus we can define the following quantities in terms of Q:

$f(u)$	$f(Q)$	$\mu_{f(Q)}(u)$	
$-u$	$-Q$ (negation)	$\mu_Q(-u)$	
λu	λQ (scalar multiplication)	$\mu_Q(u/\lambda)$,	$\lambda \neq 0$
$1/u$	$1/Q$ (reciprocal)	$\mu_Q(1/u)$,	$u \neq 0$
u^p	Q^p (power)	$\mu_Q(u^{1/p})$,	$p \neq 0$
$|u|$	$|Q|$ (absolute value)	$(Q \cup -Q) \cap [0, +\infty)$	
e^u	e^Q	$\mu_Q(\log(u))$,	$u > 0$

A fuzzy quantity is called positive (or negative) if $\inf S(Q) \geqslant 0$ (or $\leqslant 0$), where S is its support. We then write $Q \geqslant 0$ (or $Q \leqslant 0$), with strict inequality to signify that 0 is excluded from the support of Q. It is easily seen that if M is a fuzzy interval, $1/M$ is itself a fuzzy interval only if M is either negative or positive. The quantities defined in the above table extend the definitions of Moore [17].

2.2.3.2. Extension of the Four Operations (Dubois and Prade [6], Mizumoto and Tanaka [15])

Note first that the extension of any commutative (or associative) operation is itself commutative (or associative).

The sum $Q_1 \oplus Q_2$ of two fuzzy quantities is defined by

$$\mu_{Q_1 \oplus Q_2}(w) = \sup\{\min(\mu_{Q_1}(u), \mu_{Q_2}(w - u)) \,|\, u \in \mathbb{R}\} \qquad (2.23)$$

This is a sup–min version of convolution. The set $\bar{J}(\mathbb{R})$ of u.s.c. fuzzy intervals, furnished with $\oplus$, is a semigroup with neutral element 0. On the reals, $\oplus$ is the same as ordinary addition. $-Q$ is not in general the same as the group-inverse of M for $\oplus$ since $(-Q) \oplus Q$ is a fuzzy set with modal value 0, and not simply 0 itself. $(\bar{J}(\mathbb{R}), \oplus)$ is a subsemigroup of $([0, 1]^{\mathbb{R}}, \oplus)$.

The upper and lower means $E_*(Q)$ and $E^*(Q)$ [cf. (2.2) and (2.3)] satisfy a remarkable additivity property, proved for u.s.c. fuzzy intervals in [11, 26]:

$$E^*(Q_1 \oplus Q_2) = E^*(Q_1) + E^*(Q_2)$$
$$ \qquad (2.24)$$
$$E_*(Q_1 \oplus Q_2) = E_*(Q_1) + E_*(Q_2)$$

The equalities (2.24) are in contrast to the fact that if Q_1 and Q_2, considered as sets of probability measures in the sense of (1.30), are combined by convolution in the ordinary way, the additivity of the expectations is weakened to inequality for the upper and lower expectations (Dempster [1]).

The subtraction $Q_1 \ominus Q_2$ of two fuzzy quantities is defined by

$$\mu_{Q_1 \ominus Q_2}(w) = \sup\{\min(\mu_{Q_1}(w + u), \mu_{Q_2}(u)) \mid u \in \mathbb{R}\} \qquad (2.25)$$

It can of course be verified that $Q_1 \ominus Q_2 = Q_1 \oplus (-Q_2)$.

The product $Q_1 \odot Q_2$ of two fuzzy quantities is defined by

$$\mu_{Q_1 \odot Q_2}(w) = \sup\{\min(\mu_{Q_1}(u), \mu_{Q_2}(w/u) \mid u \in \mathbb{R} - \{0\})\} \qquad \text{if } w \neq 0$$

$$= \max(\mu_{Q_1}(0), \mu_{Q_2}(0)) \qquad \text{if } w = 0 \qquad (2.26)$$

The set $\bar{J}(\mathbb{R}^+)$ of u.s.c. positive fuzzy intervals, furnished with $\odot$, is a semigroup with neutral element 1. Over the reals, $\odot$ is the same as ordinary multiplication. $1/Q$ is not a group inverse since $Q \odot 1/Q$ is a fuzzy subset with modal value 1, and not simply 1 itself. 0 is the absorbing element for $\odot$. Finally, it can be verified that

$$Q_1 \odot Q_2 = (-Q_1) \odot (-Q_2)$$
$$(-Q_1) \odot Q_2 = Q_1 \odot (-Q_2) = -(Q_1 \odot Q_2)$$

Distributivity of $\odot$ over $\oplus$ does not in general hold. Nonetheless a weakened version does hold (Dubois and Prade [7]):

$$Q_1 \odot (Q_2 \oplus Q_3) \subseteq (Q_1 \odot Q_2) \oplus (Q_1 \odot Q_3) \qquad (2.27)$$

Distributivity does hold in at least the following cases:

- Q_1 is a real number.
- Q_1, Q_2, and Q_3 are u.s.c. fuzzy intervals and Q_2, Q_3 are either both positive or both negative.
- Q_1, Q_2, and Q_3 are u.s.c. fuzzy intervals, and Q_2, Q_3 are both symmetric fuzzy intervals ($Q_2 = -Q_2$, $Q_3 = -Q_3$).

These results can be proved by means of (2.13) and generalize the classic results for intervals (Moore [17]).

The quotient $Q_1 \oslash Q_2$ of two fuzzy quantities is defined by

$$\mu_{Q_1 \oslash Q_2}(w) = \sup\{\min(\mu_{Q_1}(u \cdot w), \mu_{Q_2}(u)) \mid u \in \mathbb{R}\} \qquad (2.28)$$

We of course have that $Q_1 \oslash Q_2 = Q_1 \odot 1/Q_2$. If Q_1 and Q_2 have the same sign (positive or negative) then, if Q_1 and Q_2 are fuzzy intervals, so is $Q_1 \oslash Q_2$.

2.2.3.3. Extension of max and min (Dubois and Prade [6])

It is of interest to determine whether the extension principle can give meaning to the max or the min of two intervals.

The functions $g_1(s, t) = \max(s, t)$ and $g_2(s, t) = \min(s, t)$ are isotonic on $\mathbb{R} \times \mathbb{R}$, which allows them to be straightforwardly extended since

$$\max([a, a'], [b, b']) = [\max(a, b), \max(a', b')]$$

$$\min([a, a'], [b, b']) = [\min(a, b), \min(a', b')]$$

whence the obvious construction of $\widetilde{\max}(M, N)$ and $\widetilde{\min}(M, N)$, where $\widetilde{\max}(M, N)$ and $\widetilde{\min}(M, N)$ denote the extensions of max and min in the sense of the extension principle, and M, N are u.c.s. fuzzy intervals. Note that $\widetilde{\max}(M, N)$ may be different from both M and N, as shown in Figure 2.3. $\widetilde{\max}$ and $\widetilde{\min}$ are commutative and associative, and we have

$$-\widetilde{\max}(M, N) = \widetilde{\min}(-M, -N)$$

Over the set $\tilde{J}(\mathbb{R})$ of fuzzy intervals, $\widetilde{\max}$ and $\widetilde{\min}$ are mutually distributive and satisfy the following:

$$\widetilde{\min}(M, N) \oplus \widetilde{\max}(M, N) = M \oplus N$$

$$M \oplus \widetilde{\min}(N, P) = \widetilde{\min}(M \oplus N, M \oplus P)$$

$$M \oplus \widetilde{\max}(N, P) = \widetilde{\max}(M \oplus N, M \oplus P)$$

$$\widetilde{\max}(M, N) = M \quad \text{if and only if} \quad \widetilde{\min}(M, N) = N$$

$$\widetilde{\max}(M, M) = M \qquad \widetilde{\min}(M, M) = M$$

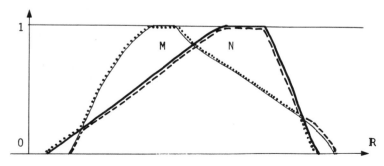

Figure 2.3. Extended minimum and maximum. $---$, $\widetilde{\max}(M, N)$; $\cdots$, $\widetilde{\min}(M, N)$.

2.2.4. The Problem of Equivalent Representations of a Function

There is always more than one mathematical representation of a function of real variables. Let f and g be two functions such that $f = g$ over $\mathbb{R}$. It is easy to see that in general $f \neq g$ when their arguments becomes fuzzy quantities. We have met this case in connection with the distributivity of the product operation over the sum of fuzzy intervals. Moore [17] points it out in connection with interval calculus.

For example, suppose f is a function of two variables x and y, and g a function of four variables x_1, x_2, x_3, x_4 such that $\forall x, y,\ f(x, y) = g(x, x, x, y)$. In general we have the inclusion

$$f(Q_1, Q_2) \subseteq g(Q_1, Q_1, Q_1, Q_2) \tag{2.29}$$

and not equality. An immediate instance of this situation is obtained by taking $g(x_1, x_2, x_3, x_4) = f(x_1, x_4) - x_2 + x_3$ since $f(Q_1, Q_2) \ominus Q_1 \oplus Q_1 \supset f(Q_1, Q_2)$ because $Q_1 \ominus Q_1 \neq 0$. g is called an *improper* representation of f insofar as the same variable occurs several times in the expression g, whereas it occurs but once in the expression f. The inclusion (2.29) is generally the case for improper representations (Dubois [4]).

The expression $g(Q_1, Q_1, Q_1, Q_2)$ conceals a certain ambiguity in that distinct variables x_1, x_2, x_3 can very well have the same fuzzy domain Q_1 of possible values. In short, different fuzzy domains necessarily correspond to different variables, but the converse is false.

Note that the result obtained from an improper representation of a function is not, strictly speaking, false; in fact, (2.29) shows that it gives a bracketing of the true result.

For real functions of real variables, it can be shown (Dubois [4]) that, with u.s.c. fuzzy intervals, equality obtains in (2.29) when g is an isotonic improper representation of an isotonic function f. For example, $f(x_1, x_4) - x_2 + x_3$ is not isotonic, even if f is. On the other hand, over the positive reals $f(x_1, x_2, x_3) = x_1(x_2 + x_3)$ and $g(x_1, x_2, x_3, x_4) = x_1 x_2 + x_3 x_4$ are isotonic, which explains why, with positive fuzzy intervals, the distributivity of the product $\odot$ over the sum $\oplus$ is preserved.

Finally, observe that, for u.s.c. fuzzy intervals, $M \oplus M = 2M$ always holds, while $M \odot M = M^2$ holds if and only if either M is positive or M is negative.

In fact, $S(M^2) \subseteq \mathbb{R}^+$ while, if $M = M_+ \cup M_-$ where $S(M_-) \subseteq \mathbb{R}^-$ and $S(M_+) \subseteq \mathbb{R}^+$, then

$$M \odot M = M_+^2 \cup M_-^2 \cup (M^+ \odot M^-)$$
$$= M^2 \cup (M^+ \odot M^-)$$

and

$$S(M^+ \odot M^-) \subseteq \mathbb{R}^-$$

N.B. $Q \oplus Q = 2Q$ does not in general hold for multimodal fuzzy quantities. For example, $2\{0, 1\} = \{0, 2\}$, but $\{0, 1\} \oplus \{0, 1\} = \{0, 1, 2\}$. □

In conclusion, we observe that if in a representation of f, the arguments x_1 and x_2 appear only once, then the representation will yield the correct membership function of $f(Q_1, Q_2)$. Such a representation does not always exist: take for example the function $f(x, y, z) = xy + yz + zx$, where each argument appears twice! Nonetheless, two isotonic representations will always turn out to be equivalent for u.s.c. fuzzy intervals.

2.3. Practical Calculation with Fuzzy Intervals

In this section we show how, in many cases, calculations with fuzzy intervals can be done as calculations on parametric representations, often without introducing any approximation.

2.3.1. Parametric Representation of a Fuzzy Interval

A good parametric representation of a fuzzy interval can be obtained with a pair of members of a class of functions $\mathbb{R}^+ \rightarrow [0, 1]$, denoted by L and R, and such that L (and also R) is decreasing in the wide sense, is u.s.c., and $L(0) = 1$; we shall also suppose that

$$\forall\, u > 0, \qquad L(u) < 1$$

$$\forall\, u < 1, \qquad L(u) > 0$$

$$L(1) = 0 \quad \text{or} \quad [L(u) > 0, \forall\, u \text{ and } L(+\infty) = 0]$$

L and R, when they satisfy these conditions, are called shape functions.

We consider u.s.c. fuzzy intervals M whose membership function can be expressed by means of two functions L and R, with four parameters $(m, \bar{m}) \in \mathbb{R}^2 \cup \{-\infty, +\infty\}$ and α, β in the form

$$\mu_M(u) = L\left(\frac{m - u}{\alpha}\right) \qquad \forall\, u \leq m$$

$$= 1 \qquad \forall\, u \in [m, \bar{m}] \qquad\qquad (2.30)$$

$$= R\left(\frac{u - \bar{m}}{\beta}\right) \qquad \forall\, u \geq \bar{m}$$

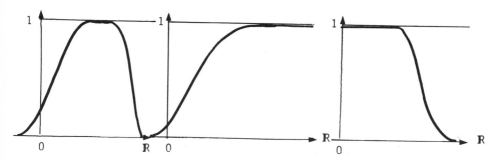

Figure 2.4. Shapes of fuzzy intervals.

This is a very general class of fuzzy intervals, since it includes the normalized u.s.c. fuzzy intervals with compact support (and therefore bounded), and, more generally, those whose membership functions satisfy the conditions

$$\lim_{u \to +\infty} \mu_M(u) \in \{0, 1\}, \qquad \lim_{u \to -\infty} \mu_M(u) \in \{0, 1\}, \qquad \dot{M} \neq \varnothing$$

Figure 2.4 shows the three possible forms of fuzzy interval: "bell-shaped" curves, nondecreasing functions, and nonincreasing functions. $[\underline{m}, \bar{m}]$ is the core of M, and $\underline{m}$ and $\bar{m}$ are, respectively, called the lower and upper modal values of m. $[\underline{m} - \alpha, \bar{m} + \beta]$ is the support of M if M is of bounded support. α and β are called the left-hand spread and the right-hand spread. Without ambiguity we may write

$$M = (\underline{m}, \bar{m}, \alpha, \beta)_{LR} \tag{2.31}$$

and we say "M is of type $L - R$". For example:

- M is a real number $m \in \mathbb{R}$. Then, by convention,

$$M = (m, m, 0, 0)_{LR}, \forall L, \forall R$$

- $M = [a, b]$. Then, by convention,

$$M = (a, b, 0, 0)_{LR}, \forall L, \forall R$$

- M is of trapezoidal form. Then (2.31) applies, with

$$L(u) = R(u) = \max(0, 1 - u)$$

- M is such that $\forall u \geq m$, $\mu_M(u) = 1$. Then

$$M = (m, +\infty, \alpha, +\infty)_{LR}, \forall R$$

- M is a fuzzy number. Then $m = \bar{m} = m$, and we write

$$M = (m, m, \alpha, \beta)_{LR} = (m, \alpha, \beta)_{LR}$$

One can choose $L(u) = \max(0, 1 - u)^p$ or $\max(0, 1 - u^p)$ with $p > 0$, or e^{-u}, e^{-u^2}, etc.

2.3.2. Exact Practical Calculation with the Four Arithmetic Operations (Dubois and Prade [6])

If $M = (m, \bar{m}, \alpha, \beta)_{LR}$ and $N = (n, \bar{n}, \gamma, \delta)_{LR}$ then the solution of the equation for $f(M, N)$ [i.e., (2.18)] reduces, for isotonic f, to finding $\lambda \in]0, 1]$ such that

$$w = f(m - \alpha L^{-1}(\lambda), n - \gamma L^{-1}(\lambda)) \qquad \text{if } w \leq f(m, n)$$
$$w = f(\bar{m} + \beta R^{-1}(\lambda), \bar{n} + \delta R^{-1}(\lambda)) \qquad \text{if } w \geq f(\bar{m}, \bar{n})$$

We are led to two equations for $L^{-1}(\lambda)$ and $R^{-1}(\lambda)$, respectively, whose complexity depends only on the complexity of an equation in u of the form

$$x = f(a + bu, c + du) \tag{2.32}$$

In particular:

- For $f(u, v) = u + v$ (isotonic on $\mathbb{R}^2$) we obtain equations of the first degree if M and N are of type LR, whence

$$M \oplus N = (m + n, \bar{m} + \bar{n}, \alpha + \gamma, \beta + \delta)_{LR} \tag{2.33}$$

- For $f(u, v) = u - v$ (hybrid on $\mathbb{R} \times \mathbb{R}$) we obtain equations of the first degree if M and N are of opposite type (LR and RL, respectively) so that

$$M \ominus N = (m - \bar{n}, \bar{m} - n, \alpha + \delta, \beta + \gamma)_{LR} \tag{2.34}$$

Thus we have a rather strong invariance theorem for the sum of fuzzy numbers: if $\tilde{J}_{LR}(\mathbb{R})$ is the class of fuzzy intervals obtained from L and R

by horizontal affine transformations followed by horizontal translations [according to (2.30), i.e., $(x, y) \mapsto (Ax + B, y)$], that is to say the set of fuzzy intervals of type $L - R$, then

Proposition 2.4

$$M, N \in \tilde{J}_{LR}(\mathbb{R}) \Rightarrow M \oplus N \in \tilde{J}_{LR}(\mathbb{R})$$

$$M \in \tilde{J}_{LR}(\mathbb{R}), \qquad N \in \tilde{J}_{RL}(\mathbb{R}) \Rightarrow M \ominus N \in \tilde{J}_{LR}(\mathbb{R})$$

This invariance property is much stronger than the corresponding property for sums of random variables, where only certain distributions (the "stable" distributions, such as the Gaussian) are invariant. This invariance theorem was independently found by Nahmias [18] for special forms of distribution.

The distributions obtained in the probabilistic context are easily compared with those obtained in the fuzzy context, in the Gaussian case.

- The sum of two normally distributed random variables with means m and n, and standard deviations σ_m and σ_n, is a normally distributed random variable with mean $m + n$ and standard deviation $(\sigma_m^2 + \sigma_n^2)^{1/2}$.
- The sum of two "Gaussian" fuzzy numbers $[L(u) = R(u) = e^{-u^2}]$ whose modal values are m and n, and whose spreads are $\alpha = \beta = \sigma_m$, $\gamma = \delta = \sigma_n$, is a Gaussian fuzzy number whose modal value is $m + n$, with left-hand and right-hand spread both equal to $\sigma_m + \sigma_n > (\sigma_m^2 + \sigma_n^2)^{1/2}$, in general.

Remark. Addition of fuzzy intervals allows a convenient representation of fuzzy proximity relations between real numbers. Let P be a fuzzy relation defined by

$$\mu_P(u, v) = 1 \qquad \text{if } |u - v| \leq \varepsilon$$

$$= L\left(\frac{|u - v| - \varepsilon}{\eta}\right) \qquad \text{if } |u - v| \geq \varepsilon$$

The fuzzy set of numbers close to M in the sense of P, where $M = (\underline{m}, \bar{m}, \alpha, \beta)_{LL}$, is defined by the composition sup–min (2.12), whose value is $M \circ P$, whatever P may be. The practical calculation of $M \circ P$ is very simple when P is defined as above, since

$$M \circ P = M \oplus E, \qquad \text{where } E = (-\varepsilon, +\varepsilon, \eta, \eta)_{LL}$$

$$= (\underline{m} - \varepsilon, \bar{m} + \varepsilon, \alpha + \eta, \beta + \eta)_{LL}$$

This result is easily extended when $L \neq R$, if P is appropriately modified.

$\square$

Proposition 2.4 does not of course hold for product or quotient; nonetheless, (2.32) remains straightforward, and

- When $f(u, v) = u \cdot v$ [isotonic over $(\mathbb{R}^+)^2$], $M > 0$, and $N > 0$, we get an equation of the second degree if M and N are of type $L - R$.

For example, for $w \leqslant \underline{m} \cdot \underline{n}$, we obtain

$$\mu_{M \odot N}(w) = L\left(\frac{\underline{n}\alpha + \underline{m}\gamma - [(\underline{m}\gamma - \underline{n}\alpha)^2 + 4\alpha\gamma w]^{1/2}}{2\alpha\gamma}\right) \quad (2.35)$$

- For $f = u/v$ [hybrid over $(\mathbb{R}^+)^2$], $M > 0$, $N > 0$, where M is of type $L - R$, and n of type $R - L$, then (2.32) is an equation of the first degree.

For example, if $w \leqslant \underline{m}/\bar{n}$, we obtain

$$\mu_{M \oslash N}(w) = L\left(\frac{\underline{m} - w\bar{n}}{\alpha + w\delta}\right) \quad (2.36)$$

Equations of the second degree also occur in the calculation of sums of the type $\sum_{i=1}^{n} \alpha_i M_i \odot N_i$, where the M_i and N_i are positive fuzzy intervals of type LR, and the α_i are real numbers all with the same sign.

On the other hand to obtain without approximation the membership function of $\prod_{i=1}^{p} M_i$, where the M_i are positive, it is necessary to solve equations of the pth degree.

2.3.3. Approximate Calculation of Functions of Fuzzy Intervals

When $f(M, N)$ cannot be calculated analytically, either it can be calculated point by point (cf. Section 2.2.1) or, when L and R are continuous, we can have recourse to approximate formulas which give an $L - R$ type approximation to the result.

When f is differentiable, (2.32) can, to the first order, in the neighborhood of $\lambda = 1$, be written

$$w \approx f(\underline{m}, \underline{n}) - f'_x(\underline{m}, \underline{n})u \cdot \alpha - f'_y(\underline{m}, \underline{n})u \cdot \gamma \quad \text{(in the isotonic case)}$$

where $u = L^{-1}(\lambda) > 0$ and $w \leqslant f(\underline{m}, \underline{n})$, and f'_x, f'_y are the partial derivatives of f; this is an equation of the first degree in u. Thus in the neighborhood of the core, if m and n are of type $L - R$ and f is isotonic, we have an approximation of type $L - R$ for $f(M, N)$ in the form

$$f_1(M, N) = (f(\underline{m}, \underline{n}), f(\bar{m}, \bar{n}), f'_x(\underline{m}, \underline{n})\alpha$$

$$+ f'_y(\underline{m}, \underline{n})\gamma, f'_x(\bar{m}, \bar{n})\beta + f'_y(\bar{m}, \bar{n})\delta)_{LR} \qquad (2.37)$$

(2.37) generalizes formulas given by Dubois and Prade [6] for the product and the quotient, and emphasizes once again the relationship between the calculus of fuzzy intervals and interval analysis.

It is easily seen that $\mu_{f_1(M,N)}$ is tangential to $\mu_{f(M,N)}$ at $f(\underline{m}, \underline{n})$ and at $f(\bar{m}, \bar{n})$.

Approximations to $f(M, N)$ can be constructed at any point corresponding to a cut at level λ by writing out the first terms of an expansion in the neighborhood of $L^{-1}(\lambda)$. The result is a function $f_\lambda(M, N)$. The expressions for $f_0(M, N)$ in the neighborhood of the support of $f(M, N)$ (if this is bounded) are specially worthy of attention.

Another kind of approximation consists of choosing a fuzzy interval with the same support and the same kernel as $f(M, N)$ but of type $L - R$ and, when M and N have bounded support, putting

$$\tilde{f}(M, N) = (f(\underline{m}, \underline{n}), f(\bar{m}, \bar{n}), f(\underline{m}, \underline{n})$$

$$- f(\underline{m} - \alpha, \underline{n} - \gamma), f(\bar{m} + \beta, \bar{n} + \delta) - f(\bar{m}, \bar{n})) \qquad (2.38)$$

(2.38) can be used for $f(x, y) = \min(x, y)$ or $\max(x, y)$, unlike (2.37).

In certain cases, the functions $\mu_{f_1(M,N)}$, $\mu_{f_0(M,N)}$, and $\mu_{\tilde{f}(M,N)}$ define two curvilinear triangles which enclose $\mu_{f(M,N)}$ and which allow a numerical estimation of the error incurred through the use of these approximations.

2.4. Further Calculi of Fuzzy Quantities

In Sections 2.2 and 2.3, the fuzzy quantities appearing in expressions to be evaluated were assumed to be noninteractive. When this is not the case, the extension principle has to be modified to take account of the interaction. The case of linear interaction is the subject of the first part of the present section. Moreover, the absence of interaction between variables leads, without any possible compensation, to accumulation of the imprecision of each of the variables. As stated above (cf. Section 2.2.2), in this case the calculus of fuzzy quantities is "pessimistic." The second part of this section outlines a possible "optimistic" calculus on fuzzy quantities in which compensation between the different sources of imprecision is maximal. Such a calculus arises in the solution of equations involving fuzzy quantities.

2.4.1. "Pessimistic" Calculus of Fuzzy Quantities with Interactive Variables

Previously, it was supposed in calculating the domain of variation of $f(x, y)$ that the variables x and y were noninteractive, that is, that the domain of variation of the pair (x, y) was the Cartesian product $M \times N$ defined by the min operator.

When this is not the case, there is relational dependency between x and y in the form of a domain D of the reference set $\mathbb{R}^2$ of (x, y). The extension principle becomes

$$\forall w, \quad \mu_{f(M,N;D)}(w)$$
$$= \sup\{\min(\mu_M(u), \mu_N(v)) \mid w = f(u, v), (u, v) \in D\} \quad (2.39)$$

D may possibly reduce to a functional relationship of the kind

$$\exists g, \quad g(x, y) = 0$$

N.B. Note that the calculation of $f(M, N)$ from an improper representation of f in the sense of Section 2.2.4 is a special case of (2.39). For example, the calculation of M^2 can be regarded as calculation of $M \odot M$ with the two occurrences of M being linked by the relation of equality.   □

One can also imagine the case where D is a fuzzy relation which defines more or less permissible zones for the values of the pair (x, y). In this case (2.39) is written

$$\forall w, \quad \mu_{f(M,N;D)}(w)$$
$$= \sup\{\min(\mu_M(u), \mu_N(v), \mu_D(u, v)) \mid f(u, v) = w\} \quad (2.40)$$

The study of the properties, and even more the calculation, of $f(M, N; D)$ are in general very difficult. We shall now give an example, considered interesting, of interaction in which it is possible to study and to calculate $f(M, N)$: the case of linear interaction. This case is studied in greater detail in Dubois and Prade [9, 10].

Given n variables $X_1, \ldots, X_n$, each constrained by a fuzzy interval

M_i $(i = 1, \ldots, n)$ we seek to calculate the possibility distribution of the variable $Z = a_1 X_1 + \cdots + a_n X_n$, where the a_i are real constants and it is known that the X_i are linked by the constraint $X_1 + X_2 + \cdots + X_n = 1$, which defines the domain D. Let N be the fuzzy set of possible values of Z. Then we can show (Dubois and Prade [10]) that

$$\sup \mu_N = \mu_{M_1 \oplus M_2 \oplus \cdots \oplus M_n}(1)$$

If this value is less than 1, we truncate the M_i in order to calculate N, that is, in accordance with Proposition 2.2, we replace μ_{M_i} by $\min(\mu_{M_i}, \sup \mu_N)$. When $\sup \mu_N = 1$, there is an n-tuple $(u_1, \ldots, u_n)$ belonging to the Cartesian product of the cores of the M_i which satisfies the constraint $\sum_i u_i = 1$.

- If the M_i are u.s.c. fuzzy intervals with compact support, then N is a u.s.c. fuzzy interval and N_λ can be calculated from the $M_{i\lambda}$, $\forall \lambda \in \,]0, 1]$. We can explicitly calculate the bounds of N_λ as a function of those of the $M_{i\lambda}$ $(i = 1, \ldots, n)$.

For example, if

$$0 \leqslant a_1 \leqslant a_2 \leqslant \cdots \leqslant a_n \quad \text{and} \quad M_{i\lambda} = [b_i, B_i]$$

then, $\forall \lambda \leqslant \sup \mu_N$:

$$\inf N_\lambda = \max_{k=1,n} \left(\sum_{j=1}^{k-1} B_j a_j + \left(1 - \sum_{j=1}^{k-1} B_j - \sum_{j=k+1}^{n} b_j\right) a_k + \sum_{j=k+1}^{n} b_j a_j \right)$$

$$(2.41)$$

$$\sup N_\lambda = \min_{k=1,n} \left(\sum_{j=1}^{k-1} b_j a_j + \left(1 - \sum_{j=1}^{k-1} b_j - \sum_{j=k+1}^{n} B_j\right) a_k + \sum_{j=k+1}^{n} B_j a_j \right)$$

$$(2.42)$$

- If the M_i are of type LL, then N is also of type LL if it is normalized (Dubois and Prade [9]). If $M_i = (\underline{m}_i, \bar{m}_i, \alpha_i, \beta_i)_{LL}$ then $N = (\inf \dot{N}, \sup \dot{N}, \gamma, \delta)_{LL}$, where γ and δ can be obtained by putting $\inf S(N) = \inf \dot{N} - \gamma$, $\sup S(N) = \sup \dot{N} + \delta$, $B_j = \bar{m}_j + \beta_j$ and $b_j = \underline{m}_j - \alpha_j$ in the above formulas, where $S(N)$ and $\dot{N}$ are the support and the core of N, respectively.

In the case where the variables X_i are linked by an arbitrary linear relation, a simple change of variables such as $Y_i = k_i X_i$ brings the situation back to the normalized case $\sum Y_i = 1$.

In Dubois and Prade [10], we give an algorithm for calculating N when the a_i are fuzzy intervals with distributions μ_{A_i}. We show the following:

- If M_i and A_i are positive u.s.c. fuzzy intervals with compact support, then N (the interactive sum of products of fuzzy intervals) is also a u.s.c. fuzzy interval and N_α can be obtained from the α-cuts $M_{i\alpha}$ and $A_{i\alpha}$.
- inf N_α can be obtained by applying (2.41) with $a_j = \inf A_{j\alpha}$ and sup N_α by applying (2.42) with $a_j = \sup A_{j\alpha}$.

The interest of being able to calculate N lies in the evaluation of expectations when probabilities are poorly known. The probability values are thus regarded as fuzzy intervals with support on $[0, 1]$, associated with variables X_i which must satisfy the normalization constraint for probabilities: $\sum_i X_i = 1$.

2.4.2. "Optimistic" Calculus of Fuzzy Quantities with Noninteractive Variables

Note first of all that equality, $A = B$, between two expressions involving fuzzy quantities is expressed as the identity of the membership functions corresponding to each side of the equality. The imprecision is therefore identical for each side. A permissible transfer of a fuzzy term from A to B, leading to a reduction in the imprecision of A, can only take place through an operation that simultaneously reduces the imprecision of B. But the operations on fuzzy quantities so far introduced in this chapter do not allow for compensation of errors and thus for the reduction of imprecision.

For example, the equation in X

$$X \ominus M = N \tag{2.43}$$

where M, N, and X are u.s.c. fuzzy intervals, does not have the solution $X = M \oplus N$, for, according to (2.43), X is more precise than N. This corresponds to the lack of any true inverse to addition of a fuzzy quantity.

All the same, (2.43) does on occasion have solutions. In fact, (2.43) corresponds to the functional equation in μ_X:

$$\forall v \in \mathbb{R}, \quad \sup\{\min(\mu_X(w), \mu_M(u)) \mid w - u = v\} = \mu_N(v)$$

When it exists, let $\mu_N(v) \wedge^{-1} \mu_M(u)$ denote the maximal solution of the equation $\min(\mu_X(w), \mu_M(u)) = \mu_N(v)$; it is equal to

$$\mu_N(v) \wedge^{-1} \mu_M(u) = \begin{cases} 1 & \text{if } \mu_M(u) = \mu_N(v) \\ \mu_N(v) & \text{if } \mu_M(u) > \mu_N(v) \end{cases}$$

For all v, we must have $\mu_X(w) \leqslant \mu_N(v) \wedge^{-1} \mu_M(w - v)$.

Therefore the maximal solution in the sense of inclusion (1.43) of equation (2.43), when it exists, is written as follows (Sanchez [22], Dubois and Prade [12]):

$$\mu_X(w) = \inf_{v \in \mathbb{R}} \mu_N(v) \wedge^{-1} \mu_M(w - v) \tag{2.44}$$

In the case where M and N are u.s.c. fuzzy *numbers* of types RL and LR, respectively, (2.44) admits of a simple interpretation. Using (2.34), it is easy to see that

$$X = (m + n, \gamma - \beta, \delta - \alpha)_{LR} \tag{2.45}$$

where $M = (m, \alpha, \beta)_{RL}$ and $N = (n, \gamma, \delta)_{LR}$. In this case the solution exists whenever $\gamma \geqslant \beta$ and $\delta \geqslant \alpha$, and it is then unique. Equation (2.45) defines an addition for fuzzy numbers where there is maximal compensation for imprecision, which is reflected in the fact that the spreads contract instead of expanding. Such "optimistic" operations are more systematically studied in Sanchez [22] and Dubois and Prade [12] for addition, and in Dubois and Prade [13] in a wider context which includes multiplication.

As they stand, equations like (2.43) have limited practical interest owing to the overrigid nature of the equality they incorporate, remembering that values of membership functions are not always known with precision. Equation (2.43) could be weakened if the equality were replaced by fuzzy-set inclusion, which would still be somewhat rigid; or, better, if an index of inclusion were used such as the one proposed in (1.66). This last approach will be used in Chapter 3 for comparison of two fuzzy quantities.

2.5. Illustrative Examples

2.5.1. Estimation of Resources in a Budget

In setting up a budget forecast one considers various sources of finance, of which some will be imprecise at the time the estimates are made, and others will be uncertain. In our example here we shall consider four sources of finance, referred to as *A, B, C,* and *D*:

- Source *A*: certain and precise, of amount $100K.
- Source B: finance is guaranteed; its amount may range from $40K to $100K according to circumstances, but we may reasonably expect an amount of $50K–$70K.
- Source *C*: it is reasonable to suppose that finance will be forthcoming, in which case it will amount to $100K–$110K; but since the final decision has not yet been made, we cannot exclude the possibility that finance will be refused.
- Source *D*: this is very uncertain, in view of the fact that negotiations have only just begun. We may perhaps hope for $20K and in any case for not more than $30K.

The various sources of finance can be represented by fuzzy quantities, whose distributions are schematically represented in Figure 2.5. Each fuzzy quantity is here regarded as a union of trapezoidal fuzzy intervals, not necessarily normalized. Each fuzzy interval M_i is represented by a quintuple whose value is

$$M_i = (\underline{m}_i, \bar{m}_i, \alpha_i, \beta_i, h_i)$$

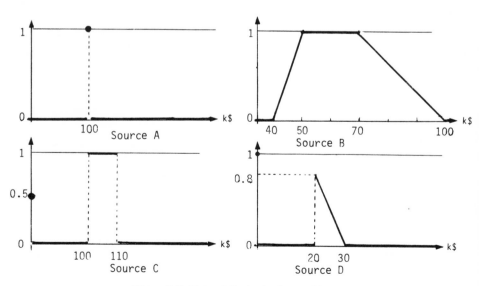

Figure 2.5. Data of the budgeting problem.

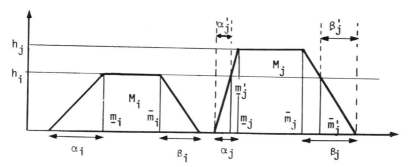

Figure 2.6. Truncation effect.

where $\underline{m}_i$ and $\bar{m}_i$ are the upper and lower modal values of M_i, α_i and β_i are its left-hand and right-hand spreads, and h_i is its height (cf. Figure 2.6). With these conventions, the four fuzzy quantities associated with the various sources of finance are represented by

$$A = (100, 100, 0, 0, 1), \qquad B = (50, 70, 10, 30, 1)$$
$$C = C_1 \cup C_2 = (0, 0, 0, 0, 0.5) \cup (100, 110, 0, 0, 1)$$
$$D = D_1 \cup D_2 = (0, 0, 0, 0, 1) \cup (20, 20, 0, 10, 0.8)$$

Applying the results of Section 3, the fuzzy quantity $M_i \oplus M_j$, where M_i and M_j are trapezoidal fuzzy intervals such as those in Figure 2.6, is a fuzzy trapezoidal interval $(\underline{m}, \bar{m}, \alpha, \beta, h)$, where

$$h = \min(h_i, h_j) \qquad \text{(truncation effect)}$$

$$\alpha = h\left(\frac{\alpha_i}{h_i} + \frac{\alpha_j}{h_j}\right), \qquad \beta = h\left(\frac{\beta_i}{h_i} + \frac{\beta_j}{h_j}\right)$$

$$\underline{m} = \underline{m}_i + \underline{m}_j - \alpha_i - \alpha_j + \alpha, \qquad \bar{m} = \bar{m}_i + \bar{m}_j + \beta_i + \beta_j - \beta$$

The sum $S = A \oplus B \oplus C \oplus D$ is obtained as the union (cf. Proposition 2.3)

$$S = \bigcup_{i=1,2} \bigcup_{j=1,2} (A \oplus B \oplus C_i \oplus D_j)$$

In the case of this example we obtain

$$S = (250, 280, 10, 30, 1) \cup (145, 185, 5, 15, 0.5)$$
$$\cup (165, 209, 5, 21, 0.5) \cup (268, 306, 8, 34, 0.8)$$

This result is displayed in Figure 2.7.

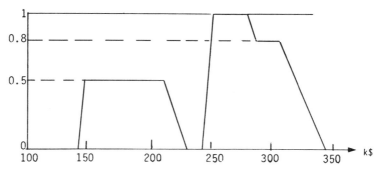

Figure 2.7. Result of the budgeting problem.

These results reveal a zone of likely finance around \$250K–\$280K. To exceed \$280K is not impossible though it is less likely (level 0.8). It does not seem very credible that receipts would only amount to \$150K–\$200K (level 0.5). In any case, they cannot go below \$140K nor go above \$340K. Note that the levels 0.5 and 0.8 which figure in the sources C and D are somewhat arbitrary, but do indicate that we are more likely (0.8) to get \$20K from source D than to get nothing from source C (0.5). These values are not combined, and reappear in the final result. It is therefore not crucial to know them precisely. What is important is that they should be distinct, so as to serve as "markers" to aid in interpretation of the final result.

The Appendix gives a BASIC program which carries out the addition of several "multitrapezoidal" fuzzy quantities such as in the above example. However, the final operation of union has not been programmed.

2.5.2. Calculation of a PERT Analysis with Fuzzy Duration Estimates

We shall consider a classical problem of organizing a project that has been broken down into tasks. The set of tasks is represented by the edges of a directed graph with no cycle; these edges also express the relations of priority among the tasks; each edge is given a value d_{ij} equal to the duration of the associated task. We know the earliest starting date t_0 of the project and possibly also its latest finishing date T_w. Thus we shall use here a "potential-stages" representation (cf. [14]).

There are situations where the durations of the tasks are poorly known a priori, and estimated subjectively. It is natural to represent these durations as fuzzy intervals; we shall denote by D_{ij} the fuzzy duration of the task corresponding to the edge (i, j) of the graph (S, A), where i and j are nodes of S and A is its set of edges.

To calculate the earliest and latest starting dates of the tasks, we generally proceed as follows in the classical case:

- Number the nodes in increasing rank order (see [14]) for a numbering algorithm).
- Obtain the earliest starting date for the group of tasks of origin i by means of the following formula:

$$t_i = \max\{t_j + d_{ji} \mid j \in P_i\}, \qquad P_i \neq \varnothing$$
$$= t_0 \qquad \text{if } P_i = \varnothing$$

where $P_i = \{j, (j, i) \in S\}$ = the set of predecessors of i.

The ranking of the nodes ensures that $\forall j \in P_i \neq \varnothing$, $j < i$. The earliest finishing date for the project is

$$t_w = \max\{t_i \mid i \in S\}$$

In the same way we obtain the latest starting date for the group of tasks with origin i as

$$T_i = \min\{T_j - d_{ij} \mid j \in S_i\} \qquad \text{if } S_i \neq \varnothing$$
$$= T_w \qquad \text{if } S_i = \varnothing$$

where $S_i = \{j \mid (i, j) \in S\}$ is the set of successors of i. The ranking of the nodes ensures that $\forall j \in S_i$, $j > i$. The interval $[t_i, T_i]$ defines for each node a *slack time* during which the tasks originating at i may start, and those ending at i may finish, without going beyond the latest finishing date. If $T_i = t_i$, then i is on a critical path, i.e., one made up of critical tasks. If $T_w \geq t_w$, where t_w is the earliest finishing date of the project, then for all i, $T_i \geq t_i$. When the tasks have imprecise durations, represented by fuzzy intervals, this algorithm is still good, if the operations of addition, subtraction, maximization, and minimization are replaced by their extensions to fuzzy arguments (Sections 2.3.2 and 2.3.3); see [4] and [5].

The program given in the Appendix contains procedures that carry out these operations on fuzzy intervals of type LR. For maximization and minimization we use the global approximation suggested by formula (2.38). If $M = (m, \bar{m}, \alpha, \beta)_{LR}$ and $N = (n, \bar{n}, \gamma, \delta)_{LR}$, then

$$\widetilde{\max}\,(M, N) = (\max(m, n), \max(\bar{m}, \bar{n}), \max(m, n)$$
$$- \max(m - \alpha, n - \gamma), \max(\bar{m} + \beta, \bar{n} + \delta) - \max(\bar{m}, \bar{n}))_{LR}$$

This approximate operation is associative.

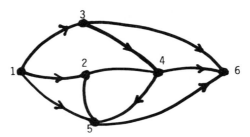

Figure 2.8. A PERT graph.

$\widetilde{\min}$ (M, N) can be obtained in the same way on replacing max by min in the above expression. In practice we use trapezoidal fuzzy intervals. Direct generalization of the algorithm for calculating the earliest and latest dates is possible because the results are monotonic functions of the data; thus we avoid the problems associated with improper representations of these results (cf. Section 2.2.4).

Consider the example in Figure 2.8 with the following fuzzy durations:

i, j	D_{ij}
1, 2	$(2, 3, 0, 1)$
1, 3	$(3, 3, 1, 3)$
1, 5	$(3, 4, 1, 1)$
2, 4	$(2, 4, 0, 1)$
2, 5	$(4, 5, 2, 3)$
3, 4	$(1, 2, 0, 0)$
3, 6	$(8, 11, 1, 4)$
4, 5	$(3, 3, 1, 2)$
4, 6	$(3, 4, 0, 2)$
5, 6	$(1, 1, 0, 1)$

Setting the earliest starting date of the project as $t_0 = (1, 1, 1, 1)$, and the latest finishing date as $T_w = (20, 21, 1, 0)$, we easily obtain the following results:

Node	Earliest starting date	Latest finishing date
1	$(1, 1, 1, 1)$	$(6, 10, 8, 2)$
2	$(3, 4, 1, 2)$	$(12, 15, 5, 1)$
3	$(4, 4, 2, 4)$	$(9, 13, 5, 1)$
4	$(5, 8, 1, 3)$	$(16, 17, 4, 1)$
5	$(8, 11, 2, 5)$	$(19, 20, 2, 0)$
6	$(12, 15, 3, 8)$	$(20, 21, 1, 0)$

For example,

$$t_4 = \widetilde{\max} (t_2 \oplus d_{24}, t_3 \oplus d_{34})$$
$$= \widetilde{\max} ((3, 4, 1, 2) \oplus (2, 4, 0, 1), (4, 4, 2, 4) \oplus (1, 2, 0, 0))$$
$$= \widetilde{\max} ((5, 8, 1, 3), (5, 6, 2, 4))$$
$$= (5, 8, 5 - \max(4, 3), \max(11, 10) - \max(8, 6))$$
$$= (5, 8, 1, 3)$$

The interval $[t_i, T_i]$ becomes a fuzzy interval bounded by fuzzy quantities: the criticality of the node i becomes more or less uncertain according to how the fuzzy intervals t_i and T_i overlap. Such an overlap can be seen for node 1, for example. See Chapter 3 for a discussion of path criticality (and also Chapter 7 of [4]). The Appendix contains a program for calculating the distributions of t_i, T_i in the form of trapezoidal fuzzy intervals.

Remarks. (1) When the data are precise, the nodes of the critical paths are found by calculating the latest possible dates starting with the earliest finishing date t_w of the project. A critical node is one for which $t_i = T_i$. This procedure is no longer meaningful with imprecise data, since t_w reflects the imprecision of the data D_{ij}. Since this imprecision must accumulate, there is a risk of counting it twice in the course of calculating the T_i starting with $T_w = t_w$. Therefore it is necessary to define T_w independently of t_w in an imprecise environment (see [4] for a more detailed discussion).

(2) A different approach to the same problems when the data are imperfectly known is to consider the D_{ij} as random variables. However, because of the problems of dependence between the different paths of the graph, and therefore between the random variables associated with the t_i and the T_i, the algorithms for the shortest and the longest paths cannot be easily adapted to probabilistic data. It is known that probabilistic PERT is a difficult problem, while the approach based on fuzzy quantities turns out to be simpler to realize, since it is "best and worst case" type, and since, because of the results of Section 2.2.4, applying the standard algorithm to fuzzy data gives the exact possibility distributions for the t_i and T_i. See [4] for further details.

2.5.3. A Problem in the Control of a Machine Tool [2, 3]

A machining operation is generally defined as the passage of a cutting or drilling tool through the surface of a metallic part. The main control parameters are the speed of the cut and the feed, which directly determine the resultant quality of the operation. The principal constraints on these parameters are the way the tool is used, the dynamics of the machine tool, and the desired state of the surface of the part.

For an isolated machining operation, the control problem can be seen as an optimization problem in which the optimization criterion may be, for example, the cost of the operation or the quantity of metal removed [3]. Nonetheless, a skilled person can very quickly determine a range of speed and feed for a given operation which will give a certain quality of result.

The problem becomes more complicated when one has to control a group of machines or a series of synchronized work stations. We then have two additional constraints which link the controls to be maintained: productivity considerations impose a work rate to be maintained which will be the same at each station, and in addition one seeks to harmonize the rates of wear of the tools so as to plan their replacement. The expert is thus led to proceed by trial and error until he finds satisfactory settings that satisfy the constraints on production rate. In general, he is unable to deal with problems of wear. This is why we have proposed in [2] an interactive method of calculation for the control settings. The method takes as data the ranges of speed and feed given by the expert in the form of fuzzy intervals. He determines for each operation and control parameter a preferred zone to ensure the quality of the cut, and a tolerable zone within which all results must fall. Thus the core and support of the fuzzy interval are obtained. The form of the membership functions can be elucidated by carrying out a test of the appropriateness of moving away from the preferred range.

Consider a station where a single operation i is carried out. Let V_i and A_i be the fuzzy intervals specifying the speed and the feed of the cut, respectively. The duration t_i of the operation is a function of these two parameters of the form

$$t_i = \frac{K_i}{a_i v_i}$$

where K_i depends on the geometry of the part. The values $\mu_{V_i}(v_i)$ and $\mu_{A_i}(a_i)$ of the membership functions can be interpreted as measures of the quality of the cut.

We can therefore formulate the control problem under a production rate constraint as follows:

$$\text{maximize } \min(\mu_{V_i}(v_i),\ \mu_{A_i}(a_i))$$

subject to the constraint

$$\frac{K_i}{a_i v_i} = \hat{t}$$

where $\hat{t}$ is the rate of working required for the station. The choice of the min operator reflects the fact that we are dealing with the logical conjunction of objectives (cf. Chapter 3). An optimal solution (a_i^*, b_i^*) satisfies

$$\mu_{V_i}(v_i^*) = \mu_{A_i}(a_i^*) = \mu_{K_i/A_i \odot V_i}(\hat{t}) = \lambda_i^*$$

where $K_i/A_i \odot V_i$ is obtained by means of the calculus of fuzzy intervals. From this solution we can find the optimal degree of satisfaction λ_i^* such that any departure from the cutting speed will cause a deterioration in the feed, and vice versa. v_i^* and a_i^* can easily be found when λ_i^* is known. If $\lambda_i^* = 1$, denoting by $[\underline{a}_i, \bar{a}_i]$ and $[\underline{v}_i, \bar{v}_i]$ the preferred ranges of feed and of speed, there is more than one possible control; for example

$$a_i^* \in [\underline{a}_i, \bar{a}_i] \cap \left[\frac{K_i}{\hat{t} \bar{v}_i}, \frac{K_i}{\hat{t} \underline{v}_i} \right]$$

Otherwise, we have, for $A_i = (\underline{a}_i, \bar{a}_i, \alpha_i, \beta_i)_{LR}$,

$$a_i^* = \underline{a}_i - \alpha_i L(\lambda_i^*) \qquad \text{if } \hat{t} > K_i/\underline{a}_i \underline{v}_i$$
$$= \bar{a}_i + \beta_i R(\lambda_i^*) \qquad \text{if } \hat{t} < K_i/\bar{a}_i \bar{v}_i$$

If λ_i^* is too low, then we conclude that the required production rate is incompatible with the required quality.

The case of a multitool station, and the problem of planning the replacement of worn-out tools, are treated in [3].

As an example consider a boring operation for a cylinder of length $L = 200$ mm and diameter $D = 100$ mm. The boring is done in two stages:

- A roughing-out first pass with wide tolerance on the cutting conditions; for example, $A_1 = (0.4, 0.8, 0.2, 0.2)$ mm, $v_1 = (80, 90, 10, 30)$ m/min;
- The finishing pass with very strict feed characteristics and a higher speed of cutting; for example, $A_2 = (0.5, 0.5, 0.3, 0.1)$ mm and $V_2 = (110, 120, 20, 10)$ m/min.

These two operations are carried out at two stations in series at which the work rate is given by $\hat{t} = 0.9$ min. We have $K_1 = K_2 = L\pi D$ (the machined surface).

Using the exact formula for the product of two fuzzy intervals of type LR [cf. formula (35)], and noting that

$$\mu_{K_i/A_i \odot V_i}(\hat{t}) = \mu_{A_i \odot V_i}(K_i/\hat{t})$$

we obtain the following results:

- During the roughing-out:

$$\lambda_1^* = 1 \quad \text{and} \quad a_1^* \in [0.78, 0.8] \text{ mm}$$
$$v_1^* \in [87, 90] \text{ m/min}$$

provided that $a_1^* v_1^* = K_1/\hat{t}$. Thus the specifications are set out.

- In the finishing stage

$$\lambda_2^* = 0.44 \quad \text{and} \quad a_2^* \simeq 0.55 \text{ mm}$$
$$v_2^* = 125.5 \text{ m/min}$$

A program to perform these calculations is given in the Appendix.

Appendix: Computer Programs

```
REM ======================================================================
REM   1. LIBRARY OF ROUTINES WHICH PROCESS FUZZY INTERVALS
REM ======================================================================

REM ======================================================================
REM  =                        SUBROUTINE  ADD/NBF                       =
REM  =                                                                  =
REM  = Parameters X, Y, Z                                               =
REM  = Aim   : Computes the sum of fuzzy intervals indexed by X and Y and =
REM  =         assigns the result  to the variable indexed by Z.        =
REM ======================================================================
ADD/NBF:
   VMI(Z)=VMI(X)+VMI(Y)
   VMS(Z)=VMS(X)+VMS(Y)
   ALPHA(Z)=ALPHA(X)+ALPHA(Y)
   BETA(Z)=BETA(X)+BETA(Y)
RETURN
REM ======================================================================
REM  =                        SUBROUTINE  SUB/NBF                       =
REM  =                                                                  =
REM  =  Parameters X, Y, Z                                              =
REM  =  Aim : Computes the subtraction of fuzzy intervals indexed by X and =
```

```
REM =            Y and assigns the result to the variable indexed by Z.        =
REM ====================================================================================
SUB'NBF:
  X1=VMI(Y)                              ! intermediary variable which
  VMI(Z)=VMI(X)-VMS(Y)                   ! prevents side effects.
  VMS(Z)=VMS(X)-X1
  X1=ALPHA(Y)
  ALPHA(Z)=ALPHA(X)+BETA(Y)
  BETA(Z)=BETA(X)+ALPHA(Y)
RETURN
REM ====================================================================================
REM =                            SUBROUTINE MOINS/NBF                             =
REM =                                                                             =
REM =  Parameters X, Y,                                                           =
REM =  Aim : Computes the negative of the fuzzy interval indexed by X and         =
REM =        assigns the result to the variable indexed by Y.                     =
REM ====================================================================================
MOINS'NBF:
  X1=VMI(X)
  VMI(Y)=-VMS(X)
  VMS(Y)=-X1
  X1=ALPHA(X)
  ALPHA(Y)=BETA(X)
  BETA(Y)=X1
RETURN
REM ====================================================================================
REM =                            SUBROUTINE MIN/NBF                               =
REM =                                                                             =
REM = Parameters : X, Y, Z                                                        =
REM = Aim   : Applies the fuzzy minimum operation to the fuzzy intervals X and Y and assigns  =
REM = the result to the variable indexed by Z. It uses the approximation formula (2.38).      =
REM ====================================================================================
MIN'NBF:
  X1=VMI(X) MIN VMI(Y)
  X2=VMS(X) MIN VMS(Y)
  X3=(VMI(X)-ALPHA(X)) MIN (VMI(Y)-ALPHA(Y))
  X4=(VMS(X)+BETA(X)) MIN (VMS(Y)+BETA(Y))
  VMI(Z)=X1
  VMS(Z)=X2
  ALPHA(Z)=X1-X3
  BETA(Z)=X4-X2
RETURN
REM ====================================================================================
REM =                            SUBROUTINE MAX/NBF                               =
REM =                                                                             =
REM = Parameters X, Y, Z                                                          =
REM = Aim : Applies the fuzzy maximum operation to the fuzzy intervals X and Y and assigns    =
REM =       the result to the variable indexed by Z. It uses the approximation formula        =
REM =       (2.38).                                                                =
REM ====================================================================================
MAX'NBF:
  X1=VMI(X) MAX VMI(Y)
  X2=VMS(X) MAX VMS(Y)
  X3=(VMI(X)-ALPHA(X)) MAX (VMI(Y)-ALPHA(Y))
  X4=(VMS(X)+BETA(X)) MAX (VMS(Y)+BETA(Y))
  VMI(Z)=X1
  VMS(Z)=X2
  ALPHA(Z)=X1-X3
  BETA(Z)=X4-X2
RETURN
REM ====================================================================================
REM =                            SUBROUTINE INIT/NBF                              =
REM =                                                                             =
REM = Parameter : XLIBR                                                           =
REM = Aim : assigns to XLIBR the index of the available fuzzy interval            =
REM =       and updates the indexation                                            =
REM ====================================================================================
INIT'NBF:
  XLIBR=PLIBR
  PLIBR=PLIBR+1
RETURN
REM ====================================================================================
REM =                            SUBROUTINE SAISIE/NBF                            =
```

```
REM =                                                                          =
REM = Parameters : XLIBR                                                        =
REM = Aim : Reads in the datafile the values which characterize  the fuz-_
REM =         zy interval indexed by XLIBR, as specified by INIT/NBF.          =
REM ===========================================================================
SAISIE'NBF:
  XXO=XLIBR
  READ VMI(XXO)
  READ VMS(XXO)
  READ ALPHA(XXO)
  READ BETA(XXO)
RETURN
REM ===========================================================================
REM =                          SUBROUTINE AFFECT/NBF                           =
REM =                                                                          =
REM = Parameters : X, Y                                                        =
REM = Aim : Assigns the fuzzy value of the variable indexed by X to the        =
REM =        variable indexed by Y.                                            =
REM ===========================================================================
AFFECT'NBF:
  VMI(Y)=VMI(X)
  VMS(Y)=VMS(X)
  ALPHA(Y)=ALPHA(X)
  BETA(Y)=BETA(X)
RETURN


REM ===========================================================================
REM     2.  COMPUTATION WITH MULTIMODAL FUZZY QUANTITIES
REM ===========================================================================
REM
REM     Constants :
REM  ---------------
MAXIND=50

REM -------------------------------------------------------------------------
REM      Definition of the representation parameters for a fuzzy quantity    -
REM -------------------------------------------------------------------------
REM  A fuzzy quantity is defined by a six-component vector sequence
REM  accessible from the following table :
REM
DIM VMI(MAXIND)                          ! lower modal value
DIM VMS(MAXIND)                          ! upper modal value
DIM ALPHA(MAXIND)                        ! left spread
DIM BETA(MAXIND)                         ! right spread
DIM HAUT(MAXIND)                         ! heights of modal values
DIM SUITE(MAXIND)                        ! a pointer to another vector if the fuzzy
REM                                      ! quantity is the union of several vectors. It
REM                                      ! gives the address of the following vector if
REM                                      ! any, or gives 0 otherwise.
REM -------------------------------------------------------------------------
REM
REM Initialization of the vector pointer :
REM  ------------------------------------
PLIBR=1

...
REM ===========================================================================
REM =                          SUBROUTINE PLUS/NBF                             =
REM =                                                                          =
REM = Parameters : X, Y, Z                                                     =
REM = Aim : Performs the sum of two fuzzy quantities addressed by X and Y      =
REM =        and assigns the result to the variable indexed by Z. These        =
REM =        fuzzy quantities can be multimodal.                               =
REM ===========================================================================
PLUS'NBF:
XO=X : YO=Y :ZO=Z
ADD'TRP:                                              ! sum of two trapezoids
  HAUT(ZO)= HAUT(XO) MIN HAUT(YO)
  ALPHA(ZO)=(ALPHA(XO)/HAUT(XO) + ALPHA(YO)/HAUT(YO))*HAUT(ZO)
  BETA(ZO)=(BETA(XO)/HAUT(XO) + BETA(YO)/HAUT(YO))*HAUT(ZO)
  VMI(ZO)=VMI(XO)+VMI(YO)-ALPHA(XO)-ALPHA(YO)+ALPHA(ZO)
  VMS(ZO)=VMS(XO)+VMS(YO)+BETA(XO)+BETA(YO)-BETA(ZO)
  IF SUITE(YO)=0 THEN GOTO SUITE'X
```

```
        YO=SUITE(YO)
        GOSUB INIT'NBF
        SUITE(ZO)=XLIBR
        ZO=XLIBR
        GOTO ADD'TRP
    SUITE'X:
        YO=Y
        IF SUITE(XO)=0 THEN SUITE(ZO)=0 : RETURN
          XO=SUITE(XO)
          GOSUB INIT'NBF
          SUITE(ZO)=XLIBR
          ZO=XLIBR
          GOTO ADD'TRP
REM =====================================================================

REM =====================================================================
REM  3. COMPUTATION OF EARLIEST STARTING AND LATEST ENDING DATES IN A PERT-GRAPH
REM =====================================================================

REM  Constants :
REM  ------------
    MAXIND=50

REM  ----------------------------------------------------------------------
REM    Definition of representation parameters of the fuzzy intervals
REM  ----------------------------------------------------------------------
REM    A fuzzy interval is viewed as a 4-component vector accessible via
REM    the following table:
REM
DIM VMI(MAXIND)                    ! lower modal value
DIM VMS(MAXIND)                    ! upper modal value
DIM ALPHA(MAXIND)                  ! left spread
DIM BETA(MAXIND)                   ! right spread
REM  ----------------------------------------------------------------------

REM Initialization of the vector pointer :
REM  -------------------------------------
PLIBR=1
REM

REM =====================================================================
REM =              COMPUTATION OF EARLIEST AND LATEST DATES             =
REM =                                                                   =
REM = A graph is encoded by numbering its nodes in increasing order,    =
REM = consistently with arc directions. In other words an arc cannot be =
REM = directed toward a node whose number is less than the number of the =
REM = arc origin.
REM =====================================================================
DEBUT:
OPEN #1,"GRAPHE.TXT",OUTPUT
REM  Graph acquisition
REM  ------------------
    READ NB'VERTICES
    DIM A%(NB'VERTICES, NB'VERTICES)
    READ NB'ARCS
    PRINT #1, "NUMBERS OF VERTICES : ",NB'VERTICES
    PRINT #1, "NUMBERS OF ARCS : ",NB'ARCS
    PRINT #1
    PRINT #1, "Source", "sink", "VMI", "VMS", "ALPHA" ; "     BETA"
    PRINT #1, "------", "----", "---", "---", "-----" ; "     ----"
    PRINT #1
    FOR I=1 TO NB'ARCS
      READ SD                      ! Source vertex
      READ SA                      ! Sink vertex
      GOSUB INIT'NBF
      A%(SD,SA)=XLIBR
      GOSUB SAISIE'NBF
      PRINT #1,SD,SA,VMI(XLIBR),VMS(XLIBR),ALPHA(XLIBR);TAB(70);BETA(XLIBR)
    NEXT I


REM  Initialization for date computations
REM  ------------------------------------
DIM PLUSTOT%(NB'SOMMETS),PLUSTARD%(NB'SOMMETS)
```

```
    FOR I=1 TO NB'SOMMETS
      GOSUB INIT'NBF
      PLUSTOT%(I)=XLIBR
      GOSUB INIT'NBF
      PLUSTARD%(I)=XLIBR
    NEXT I
  REM  Acquisition of starting time
  REM  ------------------------------------
    XLIBR=PLUSTOT%(1)
    GOSUB SAISIE'NBF

    X=XLIBR
    FOR I=2 TO NB'VERTICES              ! Earliest starting times are initia-
      Y= PLUSTOT%(I)                    ! lized with the project starting time.
      GOSUB AFFECT'NBF
    NEXT I

  REM Acquisition of end time
  REM  ------------------------------
    XLIBR=PLUSTARD%(NB'VERTICES)
    GOSUB SAISIE'NBF

    X=XLIBR
    FOR I=1 TO NB'VERTICES-1           ! Latest starting times are initialized
      Y=PLUSTARD%(I)                   ! with the project end time.
      GOSUB AFFECT'NBF
    NEXT I

    GOSUB INIT'NBF
    R=XLIBR                             ! intermediary variable

  REM Calculation of earliest starting times
  REM  ----------------------------------------------
  FOR I=1 TO NB'VERTICES-1
    FOR J=I+1 TO NB'VERTICES
      X=PLUSTOT%(I)
      IF A%(I,J)=0 THEN GOTO SUIVANT1
        Y=A%(I,J)
        Z=R
        GOSUB ADD'NBF
        X=PLUSTOT%(J)
        Y=R
        Z=X
        GOSUB MAX'NBF
    SUIVANT1: NEXT J
  NEXT I


  REM  Calculation of latest ending times
  REM  ----------------------------------------
  FOR I =NB'VERTICES TO 2 STEP -1
    FOR J=I-1 TO 1 STEP -1
      IF A%(J,I)=0 THEN GOTO SUIVANT2
        X=PLUSTARD%(I)
        Y=A%(J,I)
        Z=R
        GOSUB SUB'NBF
        X=PLUSTARD%(J)
        Y=R
        Z=X
        GOSUB MIN'NBF
    SUIVANT2: NEXT J
  NEXT I


  REM =============================================================================
  REM  4.    SUBROUTINE FOR THE COMPUTATION OF CUTTING CONDITIONS BASED ON
  REM        THE FUZZY PRODUCT
  REM =============================================================================
  REM  This program uses exact formulas for the product of two fuzzy intervals.
  REM  The example deals with the tuning of cutting parameters for the boring of a
  REM  cylinder, under a processing time constraint timp. L and D denote the length
  REM  and the diameter of the cylinder.
  REM
```

```
REM ============================================================================
pi=3.14
DIM VMI(2)
DIM VMS(2)
DIM ALPHA(2)
DIM BETA(2)
A=1                              ! cutting feed
V=2                              ! cutting speed

...

PARAM'OPTI:                      ! Subroutine

REM  Computation of the level of satisfaction if a*v is equal to (L*D*p /t).
REM  --------------------------------------------------------------------- i ------

x=L*pi*D/t
P1=(VMI(A)-ALPHA(A))*(VMI(V)-ALPHA(V))
P2=(VMS(A)+BETA(A))*(VMS(V)+BETA(V))
P3=VMI(A)*VMI(V)
P4=VMS(A)*VMS(V)
if x<=P1 or x>=P2 then satisf=0 else &
if x>=P3 and x<=P4 then satisf=1 else &
if x>P1 and x<P3 then satisf=(1-(VMI(A)*ALPHA(V)+VMI(V)*ALPHA(A)+ &
  SQR((VMI(A)*ALPHA(V)-VMI(V)*ALPHA(A))^2 +4*ALPHA(A)*ALPHA(V)*x)))/(2*ALPHA(A)*ALPHA(V)))
if x>P4 and x<P2 then satisf=(1-(-VMS(A)*BETA(V)-VMS(V)*BETA(A)+ &
  SQR((VMS(A)*BETA(V)-VMS(V)*BETA(A))^2 +4*BETA(A)*BETA(V)*x)))/(2*BETA(A)*BETA(V)))
REM calculation of optimal parameters
REM ---------------------------------
if satisf=0 then print "processing time is not compatible with the fuzzy
                                              specifications ": return
print "level of satisfaction : ";satisf
if satisf=1 then &
  print "optimal feed range : ";VMI(A) MAX x/VMS(V);",";VMS(A) MIN x/VMI(V);&
  print "optimal speed range :    ";x/(VMS(A) MIN x/VMI(V));",";x/(VMI(A) MAX x/VMS(V):&
  return
if x<P3 then aopt=VMI(A)-ALPHA(A)*(1-satisf):vopt=VMI(V)-ALPHA(V)*(1-satisf)
if x>P4 then aopt=VMS(A)+BETA(A)*(1-satisf):vopt=VMS(V)+BETA(V)*(1-satisf)
print    "optimal feed : ";aopt
print    "optimal speed : ";vopt
return
```

References

1. DEMPSTER, A. P. (1967). Upper and lower probabilities induced by a multivalued mapping. *Ann. Math. Stat.*, **38**, 325–339.
2. DUBOIS, D. (1981). A fuzzy set-based method for the optimization of machining operations. *Proc. Int. Conf. Cybernetics and Society*, Atlanta, Georgia, pp. 331–334.
3. DUBOIS, D. (1987). An application of fuzzy arithmetic to the optimization of industrial machining processes. *Math. Modelling* **9**, 461–475.
4. DUBOIS, D. (1983). Modèles Mathématiques de l'Imprécis et de l'Incertain en Vue d'Applications aux Techniques d'Aide à la Décision, Thesis, University of Grenoble.
5. DUBOIS, D., and PRADE, H. (1978). Algorithmes de plus court chemin pour traiter des données floues. *RAIRO, Rech. Operat.*, **12**(2), 213–227.
6. DUBOIS, D., and PRADE, H. (1978). Operations on fuzzy numbers. *Int. J. Syst. Sci.*, **9**, 613–626.
7. DUBOIS, D., and PRADE, H. (1979). Fuzzy real algebra: Some results. *Fuzzy Sets Syst.*, **2**, 327–348.
8. DUBOIS, D., and PRADE, H. (1980). *Fuzzy Sets and Systems: Theory and Applications.* Academic, New York.

9. DUBOIS, D., and PRADE, H. (1981). Addition of interactive fuzzy numbers. *IEEE Trans. Automatic Control*, **26,** 926–936.
10. DUBOIS, D., and PRADE, H. (1982). The use of fuzzy numbers in decision analysis. In *Fuzzy Information and Decision Processes* (M. M. Gupta and E. Sanchez, eds), North-Holland, Amsterdam, pp. 309–321.
11. DUBOIS, D., and PRADE, H. (1982). Upper and lower possibilistic expectations and applications. *4th Int. Seminar on Fuzzy Set Theory*, Linz, Austria. Technical Report no. 174, LSI, Université P. Sabatier, Toulouse, France.
12. DUBOIS, D., and PRADE, H. (1983). Inverse operations for fuzzy numbers. *Proc. IFAC Symposium on Fuzzy Information, Knowledge Representation and Decision Analysis* (E. Sanchez, ed.), Marseille, Pergamon, Oxford, pp. 399–404.
13. DUBOIS, D., and PRADE, H. (1984). Fuzzy set-theoretic differences and inclusions and their use in the analysis of fuzzy equations. *Control Cybern. (Warsaw)*, **13,** 129–146.
14. GONDRAN, M., and MINOUX, M. (1979). *Graphes et Algorithmes*. Eyrolles, Paris.
15. MIZUMOTO, M., and TANAKA, K. (1979). Some properties of fuzzy numbers. In *Advances in Fuzzy Set Theory and Applications* (M. M. Gupta, R. K. Ragade, and R. R. Yager, eds.). North-Holland, Amsterdam, pp. 153–164.
16. MOORE, R. (1966). *Interval Analysis.* Prentice-Hall, Englewood Cliffs, New Jersey.
17. MOORE, R. (1979). *Methods and Applications of Interval Analysis*. SIAM Studies on Applied Mathematics, Vol. 2, Philadelphia.
18. NAHMIAS, S. (1978). Fuzzy variables. *Fuzzy Sets Syst.*, **1,** 97–111.
19. NEGOITA, C. V. (1978). *Management Applications of Systems Theory*. Birkhaüser, Basel.
20. NGUYEN, H. T. (1978). A note on the extension principle for fuzzy sets. *J. Math. Anal. Appl.*, **64**(2), 369–380.
21. PAPOULIS, A. (1965). *Probability, Random Variables and Stochastic Processes*. McGraw-Hill, New York, Chaps. 5 and 6.
22. SANCHEZ, E. (1984). Solution of fuzzy equations with extended operations. *Fuzzy Sets Syst.*, **12,** 237–248.
23. ZADEH, L. A. (1965). Fuzzy sets. *Inf. Control*, **8,** 338–353.
24. ZADEH, L. A. (1975). The concept of a linguistic variable and its application to approximate reasoning. *Inf. Sci.*, Part 1: **8,** 199–249; Part 2: **8,** 301–357; Part 3: **9,** 43–80.
25. ZADEH, L. A. (1977). Theory of fuzzy sets. In *Encyclopedia of Computer Science and Technology* (J. Belzer, A. Holzman, and A. Kent, eds.), Marcel Dekker, New York.
26. DUBOIS, D., and PRADE, H. (1987). The mean value of a fuzzy number. *Fuzzy Sets and Syst.* **24**(3), 279–300.
27. DUBOIS, D., and PRADE, H. (1987). Fuzzy numbers: An overview. In *The Analysis of Fuzzy Information*, Vol. 1: *Mathematics and Logic* (J. C. Bezdek, ed.), CRC Press, Boca Raton, Florida, pp. 3–40.
28. KAUFMANN A., and GUPTA M. M. (1986). *An Introduction to Fuzzy Arithmetic*. Van Nostrand Rheinhold, New York.
29. BANDLER W., and KOHOUT L. (1980). Semantics of implication operators and fuzzy relational products. *Int. J. Man-Machine Studies*, **12,** 89–116.

3

The Use of Fuzzy Sets for the Evaluation and Ranking of Objects

In an article that has since become a classic [4], Bellman and Zadeh proposed the theory of fuzzy sets as a conceptual framework for problems of choice with multiple criteria. The main contribution of that article was to emphasize that objectives and constraints can be represented by fuzzy sets which subsume elements of subjective preference. In this framework, the aggregation of criteria can be viewed as a problem of combining fuzzy sets by means of fuzzy set-theoretic operations. A number of articles—among which we may cite Fung and Fu [14], Yager [31, 33, 34], Zimmerman and Zysno [35, 36, 37], and Dubois and Prade [7, 11]—have been concerned with the axiomatic or practical determination of these aggregative operations. This question is the subject of the first part of this chapter, which summarizes a more detailed survey [38].

Imprecise partial evaluations can also naturally be considered by means of fuzzy quantities such as those considered in Chapter 2. The second part of this chapter will be devoted to the ranking of imprecisely evaluated objects.

3.1. A Quantitative Approach to Multiaspect Choice

Let Ω be a set of objects to be ranked according to a set $\mathscr{C}$ of criteria. The number of objects is supposed to be finite and sufficiently small to allow them to be explicitly enumerated. The partial evaluations of the objects according to each criterion will take values in easily identifiable sets. A partial objective will be considered as a fuzzy set constraining the admissible values of the associated criterion. Thus there is an implicit hypothesis that each objective defines a total ordering on Ω. In particular, we shall not here consider the case where each objective only defines a relation of fuzzy preference on the set of objects; works that approach this problem in the framework of fuzzy sets are, particularly, those of Roy [21], Siskos, Lochard and Lombard [25], and Saaty [22]. One final hypothesis will be proposed here about the independence of choice with respect to the state of the environment. The problem of choice in an uncertain and imprecisely described environment has been approached by Okuda, Tanaka, and Asai [20], Freeling [13], and Dubois and Prade [8], among others.

This section gives a panorama of operations which amount to as many possible ways of aggregating objectives, and also a methodology for determining which, in a given situation, is the objective-aggregating operation that best respects the way in which the decision maker globally evaluates his potential actions.

3.1.1. Basic Principles of the Approach

Let X_i be the domain of evaluation of the objects in the sense of the criterion $C_i \in \mathscr{C}$; these evaluations can be represented as a mapping m_i from Ω into X_i.

The objective associated with the criterion C_i will be described by a fuzzy set G_i on X_i such that $\forall x \in X_i$, $\mu_{G_i}(x)$ is a degree of compatibility between the value of x characteristic of the object, and the desire of the decision maker. In certain cases this desire can be expressed in the form of a verbal statement represented by μ_{G_i}. The core of G_i will correspond to evaluations that are completely compatible with the objective; the evaluations that fall outside the support of G_i are totally incompatible with the objective; thus the evaluations falling in the core are mutually indifferent, as are those falling outside the support. For example, suppose that Ω is a set of apartments, X is the price, and the decision maker wishes to choose an apartment "of moderate price"; the objective "moderate" can be represented as in Figure 3.1.

μ_{G_i} cannot be exactly estimated. However, the shape of the curve can give expression to certain tendencies of the decision maker. To elicit these tendencies we do not ask him to express himself on a scale of $[0, 1]$ (which is arbitrary!), but on a discrete scale with 5–7 levels according to the perceptive thresholds of the decision maker. A simple idea is to express the degrees of compatibility between objective and evaluation verbally, and to project these levels onto $[0, 1]$ as in Table 3.1 (see also Figure 3.1).

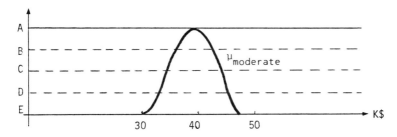

Figure 3.1. Fuzzy goal.

For the identification of μ_G we can envisage three main approaches (see the discussion in Section 1.6.1):

- Discretize X to a finite set X' and ask the decision maker to give an appreciation ($\in \{A, B, C, D, E\}$) of each evaluation $x' \in X'$, and then smooth the result.
- Represent G as a fuzzy number of type LR (cf. Chapter 2) of which the decision maker will directly provide the parameters (by fixing the limits of the core and of the support of G) and the shape (among a set of typical functions of type LR).
- Have recourse to a system of graphical representation allowing the user to trace out the curve μ_G, in this way directly visualizing the objective he seeks.

Given the objective G_i and the criterion C_i, an appraisal of each object $\omega \in \Omega$ can be made as to its compatibility with the objective, which can be described by a membership function defined by

$$\mu_i(\omega) = \mu_{G_i}(m_i(\omega)) \tag{3.1}$$

TABLE 3.1. Linguistic Scale Conventions

Degree of compatibility between objective and evaluation	Numerical convention	Verbal appreciation[a]
A Completely compatible	1	Very good
B Pretty compatible	0.75	Good
C Middling compatible	0.5	Fairly good
D Barely compatible	0.25	Poor
E Incompatible	0	Very bad

[a] Used in Section 3.1.5.

Clearly the concept of utility, as a numerical representation of preference, is very close to that of a fuzzy set such as appears in (3.1), and the membership function of a fuzzy set plays the role of a (normalized) utility function in this case. Nonetheless, a utility function always has a monetary connotation; by virtue of its abstract nature, a membership function is neutral. Again, and this is perhaps the most important point, the concept of utility function depends heavily on probability theory for the specification of the axioms that characterize it (see the book by von Neumann and Morgenstern [29]). On the other hand, the definition of membership function does not necessarily rest on probability, but rather on the existence of a relation of relative preference among the members of the reference set (see the discussion in Section 1.6.1).

We suppose that the global objective can be expressed as a hierarchy of subobjectives, at the bottom of which are q partial objectives associated with q elementary criteria C_i in terms of which we can evaluate the objectives in Ω. The global objective may perhaps be expressible as a complex verbal category whose reference set is the Cartesian product $X_1 \times \cdots \times X_q$. The fuzzy set D of objects compatible with the global objective can then be obtained by aggregation of the fuzzy sets whose membership function μ_i are defined by (3.1). We shall therefore suppose that there exists a mapping h from $[0, 1]^q$ into $[0, 1]$ such that

$$\forall \, \omega \in \Omega, \qquad \mu_D(\omega) = h(\mu_1(\omega), \ldots, \mu_q(\omega)) \qquad (3.2)$$

For the evaluation of objects we are led, therefore, to seek a fuzzy set-theoretic operation which combines the partial objectives. It will be natural to require that the operation h should satisfy the following conditions in order to be acceptable:

A1. Boundary conditions: $h(0, 0, \ldots, 0) = 0$; $h(1, 1, \ldots, 1) = 1$.
A2. $\forall \, (s_i, t_i) \in [0, 1]^2$, if $s_i \geq t_i$ then $h(s_1, \ldots, s_q) \geq h(t_1, \ldots, t_q)$.

A1 states that an action that is completely compatible with each of the objectives is a completely acceptable decision, and one that is completely incompatible with each is to be completely rejected. A2 expresses that h must not contradict the partial order defined by the vectors $(\mu_1(\omega), \ldots, \mu_q(\omega))$ on the set Ω. If $\dot{D} = \{\omega \mid \mu_D(\omega) = \sup \mu_D\}$, i.e., the set of the best objects according to the aggregation h, and if M is the set of maximal elements of the partial ordering on Ω, then we must have $\dot{D} \cap M \neq \varnothing$; when h is strictly increasing with respect to each of its arguments then we even have $\dot{D} \subseteq M$.

In what follows we shall be concerned with symmetric aggregation operators, that is to say,

A3. h is a symmetric function of its arguments.

Note that the symmetry of h does not mean that the resulting aggregation must be symmetric: in fact, we are aggregating fuzzy sets G_i, which is to say that we may very well introduce a certain kind of asymmetry by playing on the membership functions μ_{G_i}. Axiom A3 is justified when *the objectives* are of equal importance and may be interchanged in the process of aggregation. To say that the objectives are of equal importance does not mean that the criteria (i.e., the m_i) are equally critical; this we incorporate by means of the G_i. Finally, we suppose the following:

A4. h is continuous.

In [4], Bellman and Zadeh principally employ the operator min for h, which corresponds to the logical conjunction of objectives. Experimental studies have shown (cf. Thöle, Zimmermann, and Zysno [26]) that in practice this operator does not always correctly represent the attitude of an individual who aggregates verbal categories even conjunctively. Thus we are led to seek other ways of aggregating fuzzy sets.

3.1.2. Fuzzy Set-Theoretic Operations

In this section we shall axiomatically define the pointwise extensions of set-theoretic operations to fuzzy sets. These extensions are not unique, even if we can classify the operators according to the properties they conserve. In addition, there are fuzzy set-theoretic operations that do not correspond to any classical set-theoretic operation, although some of them are known in other contexts (e.g., averages).

3.1.2.1. Complementation

A pointwise complementation operator is a function c from $[0, 1]$ into $[0, 1]$ such that the complement $\bar{F}$ of the fuzzy set F is defined by

$$\forall \, \omega \in \Omega, \qquad \mu_{\bar{F}}(\omega) = c(\mu_F(\omega)) \qquad (3.3)$$

The following conditions must be imposed on c in order to preserve our intuitive idea of complementation:

C1. $c(0) = 1; c(1) = 0$ (agreement with the classical case).
C2. c is strictly decreasing (increasing the degree of F-membership by passing from ω to ω' must decrease the degree of $\bar{F}$-membership).
C3. c is an involution (double negation is affirmation).

When c is continuous, there is a unique threshold $s \in \,]0, 1[$ such that $s = c(s)$, which depends on c and in a way fixes the threshold for belonging to F when c is given. Trillas [27], assuming continuity, has solved the functional equations defining c, in the form

$$c(x) = \phi^{-1}(1 - \phi(x)) \tag{3.4}$$

where ϕ is a strictly increasing real function such that $\phi(0) = 0$ and $\phi(1) = 1$. It is easily seen that $s = \phi^{-1}(\frac{1}{2})$. When c is given, ϕ is unique. When $\phi(x) = x$ we recover classical complementation $c(x) = 1 - x$ as already presented in Section 1.5 of Chapter 1.

3.1.2.2. Union and Intersection

The operations of union and intersection can be defined pointwise by means of mappings u and i from $[0, 1]^2$ into $[0, 1]$ such that

$$\forall \,\omega \in \Omega, \qquad \mu_{F \cup G}(\omega) = u(\mu_F(\omega), \mu_G(\omega)) \tag{3.5}$$

$$\mu_{F \cap G}(\omega) = i(\mu_F(\omega), \mu_G(\omega)) \tag{3.6}$$

We shall seek to conserve as many as possible of the usual properties of union and intersection, i.e., to stay as close as possible to the Boolean lattice structure of 2^Ω. In Chapter 1 we pointed out that it is impossible to preserve this structure over the set $[0, 1]^\Omega$ of fuzzy subsets of Ω (cf. [7]). In practice, we must choose between the following two options: defining i and u so as to keep the laws of excluded middle and of noncontradiction (in which case we lose idempotence and hence also mutual distributivity for $\cup$ and $\cap$), or we keep idempotence and abandon the excluded middle and noncontradiction. We are thus led to impose only the following axioms for intersection and union:

Agreement with Classical Union and Intersection:

U0. $u(0, 1) = u(1, 1) = u(1, 0) = 1, \qquad u(0, 0) = 0.$
I0. $i(0, 1) = i(0, 0) = i(1, 0) = 0, \qquad i(1, 1) = 1.$

Commutativity:

U1. $u(x, y) = u(y, x)$.

I1. $i(x, y) = i(y, x)$.

Associativity:

U2. $u(x, u(y, z)) = u(u(x, y), z)$.

I2. $i(x, i(y, z)) = i(i(x, y), z)$.

De Morgan's Laws: There is a complementation c satisfying C1–C3, such that

U3. $c(u(x, y)) = i(c(x), c(y))$.

I3. $c(i(x, y)) = u(c(x), c(y))$.

Identity Laws:

U4. $u(x, 0) = x$ $(F \cup \emptyset = F)$.

I4. $i(x, 1) = x$ $(F \cap \Omega = F)$.

Monotonicity:

U5/I5. u and i are nondecreasing in each argument.

Continuity:

U6/I6. u and i are continuous.

U1–U4 and I1–I4 hold in the classical theory; U5 and I5 are natural because if ω belongs less to F and to G than ω' then ω' cannot belong more to $F \cup G$ nor more to $F \cap G$ than ω does. Continuity is a technical condition required when Ω is infinite.

Axioms I0, I1, I2, I4, I5 make $([0, 1], i)$ a semigroup with identity 1; the intersection operations i are called "triangular norms" in stochastic geometry because of their role in the expression of triangular inequalities (cf. Menger [19], Schweizer and Sklar [23]). The De Morgan transformation expressed by axioms U3 and I3 exchanges i and u; $([0, 1], u)$ is thus a semigroup with identity 0 known as a "co-norm." By results from the theory of functional equations (Aczél [1], Ling [18]), the main classes of intersections and of unions can be characterized as follows:

The Idempotent Operations

$$i(x, y) = \min(x, y), \qquad u(x, y) = \max(x, y)$$

These are mutually distributive; min and max are the *sole* operators of
union and intersection that satisfy I0–I5 and U0–U5 and that are
idempotent and mutually distributive. Moreover, the min operator is the
maximal intersection operator, that is to say

$$\forall x, \quad \forall y, \qquad i(x, y) \leq \min(x, y)$$

dually,

$$\forall x, \quad \forall y, \qquad u(x, y) \geq \max(x, y)$$

The Strictly Monotonic Operators. These are operators chosen such
that

$$\forall x \in \,]0, 1[, \qquad i(x, x) < x, \qquad u(x, x) > x$$
$$\forall y' > y, \qquad i(x, y') > i(x, y), \qquad u(x, y') > u(x, y)$$

Their prototype is the product operator $(x \cdot y)$ for intersection, and the
probabilistic sum $(x + y - x \cdot y)$ for union. These two operators satisfy
De Morgan's laws with $c(x) = 1 - x$. Their general form is

$$i(x, y) = f^{-1}(f(x) + f(y)) \tag{3.7}$$

where f is a continuous decreasing one-to-one mapping from $[0, 1]$ onto
$[0, +\infty)$, such that $f(0) = +\infty$, $f(1) = 0$, and

$$u(x, y) = \phi^{-1}(\phi(x) + \phi(y)) \tag{3.8}$$

where ϕ is a continuous increasing one-to-one function from $[0, 1]$ onto
$[0, +\infty)$ such that $\phi(0) = 0$, $\phi(1) = +\infty$.

We shall call these *strict union and intersection,* from the term for the
triangular norms and conorms which model them. They are never
distributive one over the other, nor idempotent, and they never satisfy
the laws of excluded middle and of noncontradiction.

Parametrized families of strict operators have been proposed in the
literature. Hamacher [15] has studied the only strict intersections which
are rational functions of their arguments:

$$i(x, y) = \frac{xy}{\gamma + (1 - \gamma)(x + y - xy)}, \qquad \gamma \geq 0$$

The corresponding unions are obtained by duality, with $c(x) = 1 - x$.

For $\gamma = 1$ we get the product. Frank [12] has studied the only operators such that $u(x, y) + i(x, y) = x + y$. They are defined by

$$i(x, y) = \log_s \left[1 + \frac{(s^x - 1)(s^y - 1)}{s - 1} \right], \qquad s > 0$$

For $s = 0$, we obtain $\min(x, y)$; for $s = 1$, the product; finally, $\lim_{s \to \infty} i(x, y) = \max(0, x + y - 1)$, which is not a strict operator.

The Nilpotent Operators. The prototypes of these operators are

Intersection: $\quad i(x, y) = \max(0, x + y - 1)$

Union: $\qquad\quad u(x, y) = \min(1, x + y) \qquad$ (bounded sum)

These operators satisfy De Morgan's laws for $c(x) = 1 - x$.
    Their general form is, for the intersection,

$$i(x, y) = f^*(f(x) + f(y))$$

where f is a decreasing one-to-one mapping from $[0, 1]$ onto $[0, f(0)]$ such that

$$f(0) = 1, \qquad f(1) = 0, \qquad \text{and } f^*(x) = \begin{cases} f^{-1}(x) & \text{if } x \in [0, 1] \\ 0 & \text{if } x \geq 1 \end{cases}$$

and, for the union,

$$u(x, y) = \phi^*(\phi(x) + \phi(y))$$

where ϕ is a complementation generator and

$$\phi^*(x) = \begin{cases} \phi^{-1}(x) & \text{if } x \in [0, 1] \\ 1 & \text{if } x \geq 1 \end{cases}$$

The following properties of the nilpotent operators can be verified:

- There is a lack of mutual distributivity and of idempotence.
- Every complementation generator c engenders an intersection and a union that are dual in the De Morgan sense and satisfy the laws of excluded middle and noncontradiction for this complementation. u and c are generated by ϕ, and i is generated by $f = \phi(1) - \phi$.

Parametrized families of nilpotent operators have been proposed in the literature: the intersection T_p of Schweizer and Sklar [23] and that of Yager [33] which are, respectively, generated by $f(x) = 1 - x^p$ for $p > 0$, and by $f(x) = (1 - x)^q$ for $q > 0$. When $p = q = 1$, $i(x, y) = \max(0, x + y - 1)$; for $p = 0$, $i(x, y) = x \cdot y$; for $q = +\infty$, $i(x, y) = \min(x, y)$.

Intersection and union operators that are not idempotent nor strict nor nilpotent can be found, such as the family of intersections proposed by Dubois and Prade [7]:

$$i(x, y) = \frac{x \cdot y}{\max(x, y, \alpha)}, \qquad \alpha \in \,]0, 1[$$

For $\alpha = 0$, $i(x, y) = \min(x, y)$; for $\alpha = 1$, $i(x, y) = x \cdot y$.

3.1.2.3. Set-Theoretic Operators Specific to Fuzzy Set Theory

In classical set theory, there are only two ways to combine sets *symmetrically* and so that axiom A1 is satisfied, namely,

$$\omega \in A, \qquad \omega \in B \Rightarrow \omega \in A * B$$
$$\omega \in \bar{A}, \qquad \omega \in \bar{B} \Rightarrow \omega \notin A * B$$

where $*$ must be either $\cup$ or $\cap$. This is no longer so for fuzzy sets. We have seen that the fuzzy operators of intersection (or union) were less than (or greater than) the min (or max) operators. Thus we have covered only a fraction of possible aggregation operators. In this section we shall study averaging operators that lie between min and max, and also operators that are self-dual in the De Morgan sense. These two kinds of fuzzy set-theoretic operations have no classical counterpart, but are often well known, as it turns out. What is new here is situating them in a set-theoretic context.

The Averaging Operators. The mean of two fuzzy sets is defined by means of a mapping m from $[0, 1]^2$ into $[0, 1]$ such that

M1. $\min(x, y) \leqslant m(x, y) \leqslant \max(x, y)$, $\forall x, y$; but $m \notin \{\min, \max\}$.
M2. $m(x, y) = m(y, x)$.
M3. m is a nondecreasing function of each of its arguments.

We demonstrate the following remarkable properties:

1. It is easily seen that m is idempotent. Conversely, idempotence and M3 imply the inequalities in M1.
2. If m is strictly increasing then it cannot be associative [7].
3. The only associative means are the medians defined for a threshold $\alpha \in \,]0, 1[$ by the following [11]:

$$
\begin{aligned}
\text{med}(x, y, \alpha) &= y & &\text{if } x \leq y \leq \alpha \\
&= \alpha & &\text{if } x \leq \alpha \leq y \\
&= x & &\text{if } \alpha \leq x \leq y
\end{aligned}
$$

Thus associativity is incompatible with the concept of average. We use instead bisymmetry, which is written

M4. $\forall x, y, z, t, \quad m[m(x, y), m(z, t)] = m[m(x, z), m(y, t)].$

Note that associativity implies bisymmetry. The functional equations M1–M4, with the hypotheses that m is continuous and strictly increasing, have been solved (Aczel [1]). The general form of the solution is

$$
m(x, y) = k^{-1}\left(\frac{k(x) + k(y)}{2}\right) \tag{3.9}
$$

where k is continuous and strictly monotonic.

For example, $k(x) = x^\alpha$, $\alpha \in \mathbb{R}$ gives a family m_α with, as special cases,

α	$m_\alpha(x, y)$	
$-\infty$	$\min(x, y)$	
-1	$\dfrac{2xy}{x + y}$	(harmonic mean)
0	$(xy)^{1/2}$	(geometric mean)
1	$\dfrac{x + y}{2}$	(arithmetic mean)
$+\infty$	$\max(x, y)$	

and in which $\alpha_1 < \alpha_2 \Rightarrow m_{\alpha_1} < m_{\alpha_2}$.

Note that the median $\text{med}(x, y, \alpha)$ is bisymmetric but is not of the form (3.9).

The Symmetric Sums [24]. Silvert has suggested study of a semantically very interesting class of aggregation operators for fuzzy sets: those that are self-dual in the De Morgan sense, i.e., functions $\sigma : [0, 1]^2 \to [0, 1]$ which satisfy

S1. $\sigma(0, 0) = 0, \qquad \sigma(1, 1) = 1.$
S2. σ is commutative.
S3. σ is a nondecreasing function of each argument.
S4. σ is continuous.
S5. $1 - \sigma(x, y) = \sigma(1 - x, 1 - y).$

N.B. Axiom S5 can be generalized to an arbitrary complementation operator c (cf. Dubois [6, 38]). $\qquad\qquad\qquad\qquad\qquad\qquad\qquad\square$

Every sum that is symmetric in the sense of S5, and that satisfies S1–S4, can be put in the form (Silvert [24])

$$\sigma(x, y) = \frac{g(x, y)}{g(x, y) + g(1 - x, 1 - y)} \qquad (3.10)$$

where g is a continuous nondecreasing and nonnegative function such that $g(0, 0) = 0$.
It can be shown (Silvert [24]) that

- $\sigma(x, 1 - x) = \frac{1}{2}.$
- $\sigma(0, 1)$ is undefined when $g(0, x) = 0 \; \forall \, x$; otherwise $\sigma(0, 1) = \frac{1}{2}$.
- The only associative symmetric sum that is a mean in the sense of M1 is $\text{med}(x, y, \frac{1}{2})$.

Therefore we must choose between associativity and idempotence for symmetric sums.
Using the classical results for the functional equation for associativity, Dombi [5] and Dubois [6] have shown that the strictly increasing *associative* symmetric sums can be put in the form

$$\sigma(x, y) = \psi^{-1}(\psi(x) + \psi(y)) \qquad (3.11)$$

where ψ is strictly monotonic and such that $\forall \, x, \psi(1 - x) + \psi(x) = 0$, and $\psi(0)$ and $\psi(1)$ are not bounded. Thus $([0, 1], \sigma)$ is a semigroup with identity $\frac{1}{2}$, and absorbing elements 0 and 1. Further, we then have, for $x < \frac{1}{2} < y$,

$$\sigma(x, x) < x, \qquad \sigma(y, y) > y, \qquad x < \sigma(x, y) < y$$

Finally, associativity must be limited to the domain $]0, 1[$, since for associative symmetric sums $\sigma(0, 1)$ is undefined.

An example of an associative symmetric sum is generated by $\sigma(x, y) = x \cdot y$:

$$\sigma_0(x, y) = \frac{xy}{1 - x - y + 2xy}$$

On the other hand, the operation $\sigma_{\hat{+}}(x, y)$, generated by $g(x, y) = x + y - xy$:

$$\sigma_{\hat{+}}(x, y) = \frac{x + y - xy}{1 + x + y - 2xy}$$

is not associative, since $g(0, x) \neq 0$, i.e., 0 is not absorbing.

Now consider idempotent symmetric sums. Since they are nondecreasing, they are averaging operators in the sense of M1–M3. If we adjoin the properties of bisymmetry (M4) and of being strictly increasing, then we can verify that there is but one single idempotent symmetric sum which is strictly increasing and bisymmetric, namely, the arithmetic mean, generated by $g(x, y) = x + y$.

For g it is possible to choose, for example a union or intersection operation. With $g = \min$ or $\max$, we have, respectively,

$$\sigma_{\min}(x, y) = \frac{\min(x, y)}{1 - |x - y|}, \qquad \sigma_{\max}(x, y) = \frac{\max(x, y)}{1 + |x - y|}$$

These two operators are means, but are not bisymmetric. For $x + y \leq 1$, we have the inequalities

$$\frac{xy}{1 - x - y + 2xy} \leq \frac{\min(x, y)}{1 - |x - y|} \leq \frac{x + y}{2}$$

$$\leq \frac{\max(x, y)}{1 + |x - y|} \leq \frac{x + y - xy}{1 + x + y - 2xy}$$

For $x + y \geq 1$, the inequalities are reversed.

Figures 3.2–3.13 illustrate the behavior of some of the fuzzy set operations introduced in this section.

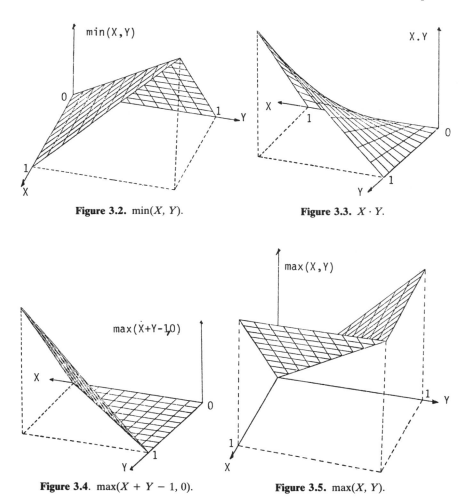

Figure 3.2. min(X, Y).   Figure 3.3. X · Y.

Figure 3.4. max(X + Y − 1, 0).   Figure 3.5. max(X, Y).

3.1.3. Application to the Combination of Criteria

3.1.3.1. The Case of Two Objectives of Equal Importance

When the objectives are defined in all-or-nothing terms, i.e., G_i is an ordinary subset of X_i (which expresses a constraint), then the only possible aggregations are the conjunction of the objectives ($D = G_1 \cap G_2$) or their disjunction (if they are interchangeable) ($D = G_1 \cup G_2$). In this case we rule out any compromise between the two criteria. When the

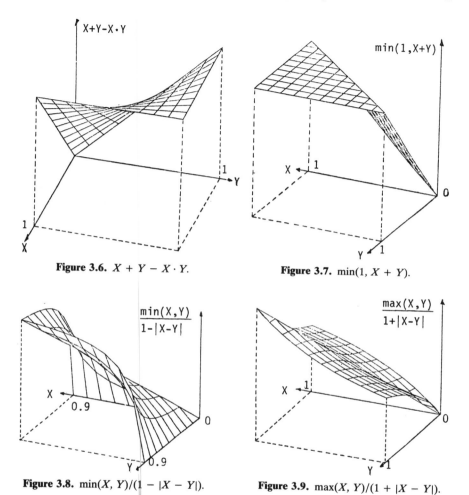

Figure 3.6. $X + Y - X \cdot Y$.

Figure 3.7. $\min(1, X + Y)$.

Figure 3.8. $\min(X, Y)/(1 - |X - Y|)$.

Figure 3.9. $\max(X, Y)/(1 + |X - Y|)$.

objectives have gradations, compromise is a natural attitude for the decision maker to adopt. Nonetheless, the two other attitudes (viz., simultaneous satisfaction of the two objectives or the redundancy of one of them) are also quite natural. We can therefore distinguish *three* fundamental attitudes for a decision maker faced with a combination of criteria.

A natural axiom for the attitude of simultaneous satisfaction of objectives is

A5. $\forall s, t, \qquad h(s, t) \leqslant \min(s, t)$

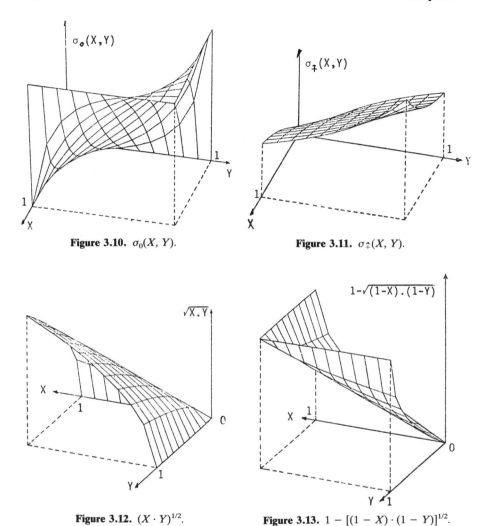

Figure 3.10. $\sigma_0(X, Y)$.

Figure 3.11. $\sigma_{\updownarrow}(X, Y)$.

Figure 3.12. $(X \cdot Y)^{1/2}$.

Figure 3.13. $1 - [(1 - X) \cdot (1 - Y)]^{1/2}$.

i.e., the global evaluation of an action cannot be better than the worst of the partial evaluations. We shall call these operators "conjunctions." An important subclass is the set of associative conjunctions which served as models for fuzzy set intersections.

   An operator expressing redundancy of objectives will satisfy an axiom dual to A5:

   A6. $\forall s, t, \qquad h(s, t) \geqslant \max(s, t)$

i.e., it is the best of the partial evaluations which governs the global evaluation. Such operators will be called "disjunctions." An important subclass is that of the associative disjunctions which served as models for fuzzy set unions.

An operator expressing compromise will satisfy the following axiom, which is complementary to A5 and A6:

A7. $\min < h < \max$

i.e., the global evaluation is intermediate between the partial evaluations. In this case we are led to averaging operators.

As well as these three "pure" attitudes, we can imagine "hybrid" attitudes, such as in the following example in which ω is an object. If ω scores badly on both criteria, it benefits from a relatively indulgent judgment (disjunction). If it scores well on both, the global judgment is severe (conjunction). This kind of aggregation is well subsumed by a symmetric sum such as the median $\text{med}(s, t, \frac{1}{2})$; on the other hand, all the associative symmetric sums (save the median) express a different hybrid attitude, namely,

$$\text{if } \max(s, t) < \tfrac{1}{2}, \qquad h(s, t) \leqslant \min(s, t)$$

$$\text{if } \min(s, t) > \tfrac{1}{2}, \qquad h(s, t) \geqslant \max(s, t)$$

$$\text{if } s \leqslant \tfrac{1}{2} \leqslant t, \qquad s \leqslant h(s, t) \leqslant t$$

which corresponds to the attitude dual to the one described above.

The interest in having functional representations for these operators is that we then have available parametrized families such as those given above, which will allow identification of the operators. A continuum of attitudes between conjunction and disjunction can be covered by a further parametrized family on putting

$$h(s, t) = i(s, t)^{\gamma} \cdot u(s, t)^{1-\gamma}, \qquad \gamma \in [0, 1]$$

where γ represents a degree of trade-off between objectives. This idea was suggested by Zimmermann and Zysno [35].

Another approach to expressing attitudes intermediate between conjunction and disjunction has been proposed by Yager [41]. The idea is to introduce fuzzy quantifiers (i.e., fuzzy sets on $[0, 1]$ expressing ill-defined proportions) into the aggregation scheme and to evaluate, for instance, to what extent "most" of the objectives are attained, where "most" is the name of a fuzzy quantifier. When "most" is replaced by "all," or "at least one," we recover, respectively, conjunction and disjunction.

Despite the apparently rich catalogue of operators presented in this chapter, it must be remembered that they cover only part (but a substantial part) of possible attitudes in the face of choice; see the example at the end of this part of the chapter (Section 3.1.5). Further research is needed to reveal a more exhaustive picture.

3.1.3.2. Aggregation of q Criteria by Symmetrical Combination of Objectives

The above operators can be extended to $q > 2$ objectives of equal importance, provided they have a structure that "naturally" lends itself to increase in the numbers of their arguments while preserving the commutativity of Axiom A3. Thus let us recursively define the operator $h^{(q)}$ with q arguments by

$$h^{(2)} = h$$
$$h^{(q)}(s_1, \ldots, s_q) = h(h^{(q-1)}(s_1, \ldots, s_{q-1}), s_q)$$

If h is associative and commutative then $h^{(q)}$ satisfies A3 and is the natural extension of h for aggregation of q criteria. This procedure can be used for conjunctions, disjunctions, associative symmetric sums, and parametrized medians.

For averaging operators other than the median, we use bisymmetry for the extension to q arguments. In general, then, we shall have

$$h(s_1, \ldots, s_q) = k^{-1}\left(\frac{1}{q}\sum_{i=1}^{q} k(s_i)\right)$$

Some (possibly nonassociative) symmetric sums also have a natural extension, which can be defined in terms of the canonical form described in Section 3.1.2.3. If the function g is associative we quite naturally have

$$h(s_1, \ldots, s_q) = \frac{g^{(q)}(s_1, \ldots, s_q)}{g^{(q)}(s_1, \ldots, s_q) + g^{(q)}(1 - s_1, \ldots, 1 - s_q)}$$

where $g^{(q)}$ is defined, like $h^{(q)}$, in terms of the generator g of the symmetric sum with two arguments. This procedure suffices to generate the n-ary forms of all the given symmetric sums.

Note that one can imagine other kinds of symmetric aggregation of objectives, involving several kinds of operators. For example, if we specify that two out of the three objectives G_1, G_2, G_3 must be attained then we are led to an aggregation of the kind (Yager, [41]):

$$H(G_1, G_2, G_3) = (G_1 \cap G_2) \cup (G_2 \cap G_3) \cup (G_1 \cap G_3)$$

3.1.3.3. Criteria of Unequal Importance

The concept of the *importance* of one criterion relative to another has received very little clarification. The meaning given to this word by different decision makers or in different situations varies considerably. Without pretending to any definitive elucidation of this very tough problem, we shall attempt to isolate some possible interpretations of the concept of importance.

Use of Thresholds of Satisfaction. One interpretation of importance of criteria could be the existence for each criterion C_i of a minimum threshold x_i^0 which must be attained before the evaluation $m_i(\omega)$ of the object ω can be considered acceptable. Here we are implicitly supposing that the object ω is to be preferred in the sense of criterion C_i to the extent given by the magnitude of $m_i(\omega)$. It is clear that this approach is a special case of the general approach adopted here, namely, fuzzy objectives, since knowing a threshold x_i^0 is equivalent to having an objective G_i such that $\mu_{G_i}(x) = 1$ for all $x \geq x_i^0$, and $= 0$ otherwise. Thus the *symmetrical* aggregation of fuzzy objectives amounts to adopting one of the possible interpretations of the concept of importance of criteria: the more important a criterion is judged to be, the more demanding and specific the corresponding objective will be, in the sense of inclusion (1.43).

Weighting of Objectives. A very common method of expressing unequal importance of criteria is to attach a weight to each one, and to incorporate this weighting into the process of aggregation. From this point of view, the convex combination

$$m_{\mathbf{P}}(\omega) = \sum_{i=1}^{q} p_i m_i(\omega), \qquad \sum_{i=1}^{q} p_i = 1 \qquad (3.12)$$

of criteria is the commonest procedure. Its main justification lies in the fact that the whole set of maximal elements of Ω can be covered by varying the weighting $\mathbf{P}$ and maximizing the criterion $m_{\mathbf{P}}(\omega)$ for each $\mathbf{P}$. Nonetheless, the a priori evaluation of the weighting is problematical, all the more so in that the m_i often refer to magnitudes that have nothing in common (e.g., the speed and the price of a car). Also, (3.12) does not apply if $m_i(\omega)$ is not a number!

Let us suppose that it is reasonable to define fuzzy objectives G_i, $i = 1, \ldots, q$, for each criterion, and that we wish to weight the objectives; then a quantity such as

$$\sum_{i=1}^{q} p_i \mu_{G_i}(m_i(\omega)) = \mu_D(\omega), \qquad \sum_{i=1}^{q} p_i = 1 \qquad (3.13)$$

is more satisfactory that (3.12) since it combines magnitudes of the same kind. The coefficient p_i measures the pertinence of the objective G_i, the extent to which it contributes to the global objective. Note that operators of the kind described by (3.13) are nonsymmetric averaging operators.

One would like to extend the notion of weighting to other kinds of aggregation. This will be possible when we have additive representations of the operators. Any operator of the form $k^{-1}(\frac{1}{2}[k(a) + k(b)])$ easily generalizes to

$$h_{\mathbf{P}}(s_1, \ldots, s_q) = k^{-1}\left(\sum_{i=1}^{q} p_i k(s_i)\right) \qquad \text{with } \sum_{i=1}^{q} p_i = 1 \qquad (3.14)$$

When the operator h is of the form $\psi^{-1}(\psi(a) + \psi(b))$ we can adopt the following approach. In symmetrical aggregation of q objectives, each criterion counts as 1. In the weighted version the weight of each criterion is left free, on condition that in aggregate the total weight equals q (i.e., the "equivalent" number of criteria is always q). Thus we can put

$$h_{\mathbf{P}}(s_1, \ldots, s_q) = \psi^{-1}\left(q \sum_{i=1}^{q} p_i \psi(s_i)\right) \qquad \text{with } \sum_{i=1}^{q} p_i = 1 \qquad (3.15)$$

Equation (3.15) can be applied immediately to most conjunction and disjunction operators, and to many symmetric sums. For the latter, (3.15) is applied to a function g generating them when this function is associative. When the aggregation operator is a product, we find the approach proposed by Yager [32] on empirical grounds, which consists of weighting the membership functions μ_{G_i} by raising them to powers $q \cdot p_i$; this amounts to modifying the objectives G_i, as was done in introducing thresholds of satisfaction. An important case left out of this picture is that of the min and max operators; this we handle by means of the possibility of a fuzzy event. Let F_ω be the fuzzy set of objectives attained by an object ω:

$$\forall \omega, \quad \forall i, \qquad \mu_{F_\omega}(G_i) = \mu_{G_i}(m_i(\omega)) \triangleq s_i \qquad (3.16)$$

Let $\pi = (\pi_1, \ldots, \pi_q)$ be a possibility distribution on the set of objectives, representing the fuzzy set of pertinent criteria.

One way to introduce a weighting into the aggregation by means of max or min is to put, respectively,

$$h_\pi(s_1, \ldots, s_q) = \min_{i=1,q} \max(s_i, 1 - \pi_i) \qquad \text{with } \max_{i=1,q} \pi_i = 1 \quad (3.17)$$

for min, and for max:

$$h_\pi(s_1, \ldots, s_q) = \max_{i=1,q} \min(s_i, \pi_i) \qquad \text{with } \max_{i=1,q} \pi_i = 1 \qquad (3.18)$$

In this case the degree of compatibility $\mu_D(\omega)$ with the global objective is expressed, according as we are dealing with $h = \min$ or $h = \max$, in the form of the necessity or the possibility, respectively, of the fuzzy event F_ω (cf. Chapter 1, Section 1.7), in the sense of the possibility distribution π. When the objectives are of equal importance (i.e., $\forall i$, $\pi_i = 1$), then (3.17) and (3.18) reduce, respectively, to aggregation by min and by max.

The aggregation scheme (3.17) was first proposed by Yager [40] with a different justification. Namely, (3.17) can be interpreted as "for each objective i, *if* i is important *then* it is attained," where the *"if–then"* statement is an implication in a multivalued logic and has degree of truth equal to $\max(s_i, 1 - \pi_i)$; $\max(y, 1 - x)$ is a possible multivalued implication connective (see Section 4.3.1 for other possibilities).

Equations (3.17) and (3.18) are also special cases of "fuzzy integral" in the sense of Sugeno (see references 29 and 6 of Chapter 1 for this notion), and can thus be interpreted as medians, as pointed out by Kandel and Byatt [6] (see also reference 6, p. 134, of Chapter 1). The relation between (3.17) and (3.18), Sugeno's integral, and its expression as a median is investigated in detail in Dubois and Prade [39].

This interpretation is especially satisfactory since the notion of the median is the ordinal counterpart of the metrical notion of the mean (which presupposes additivity). That is to say that (3.17) and (3.18) are indeed the equivalents of (3.13). Moreover, (3.13) is easily interpreted in terms of probability since $\mu_D(\omega)$ is then the probability of the fuzzy event F_ω (cf. Chapter 1, Section 1.7).

Asymmetric Specification of the Global Objective. Situations arise where the global objective is stated in a complex way in terms of partial objectives, where the latter are the bottom of a hierarchy which one may describe informally, in the terminology of artificial intelligence, as an and/or tree structure. For example, suppose the purchase of a car is conditioned by such requirements as "a very powerful car, or else a not very powerful car but cheap"; one then has a structure for the aggregation operator defined by

$$D = G_1 \cup (G_2 \cap G_3)$$

where G_1 and G_2 are defined on the same scale. An operator that subsumes this desire can be realized using the maximum and product operators as follows:

$$h(s_1, s_2, s_3) = \max(s_1, s_2 \cdot s_3).$$

It can clearly be seen that this kind of asymmetry

- Is quite different from those occurring above.
- Only occurs when the number of elementary objectives exceeds two.

In conclusion, it is chiefly in hierarchical groupings of the elementary objectives that the distinction between the three fundamental attitudes of conjunction, compromise, and redundancy has full force. It will have been observed that most of the operators encountered above are (more or less) homomorphic to addition; conjunction operators have analogs expressing compromise, etc., to the extent that we may doubt the merits of aggregating criteria by using $\min(a + b, 1)$ rather than $\frac{1}{2}(a + b)$ or $\max(0, a + b - 1)$, but for the fact that the first of these cannot distinguish between the better decisions nor the last between the worse. In fact, the importance of choosing the right operator becomes clearer when a partially aggregated evaluation is to be combined in turn with other evaluations. This quantitative approach to aggregation of criteria has relationships with multidimensional utility (see, for example, Keeney and Raiffa [17]), of which preliminary discussion can be found in [6] and [38].

3.1.4. Identification of Operators

Since we have available a wide range of operators which can express numerous different attitudes that a decision maker may have, we can envisage use of a questionnaire to seek the operation in a given list which most faithfully represents his attitudes.

For simplicity, consider the case of two objectives G_1 and G_2 of equal importance which are to be aggregated. Identification of the appropriate operator will be based on data for a restricted subset T of typical objects whose degrees of conformity with each objective are known. These objects are to have contrasting characteristics, i.e., one will be incompatible with G_1 but completely compatible with G_2, another moderately compatible with both, and so on.

We envisage two stages in the identification:

1. We take a restricted but well-spread list of typical operators. The decision maker is asked to make a global judgment for the objects in T, using a verbal scale of judgment such as the one proposed in Section 3.1.1. A very small number of well-chosen objects can suffice for this procedure, as will be seen in the example below.

2. The first stage will have determined one or more plausible operators. To refine the determination of an operator, we can choose a parametrized family of operators whose meaning corresponds to the decision maker's type of attitude. He is then questioned again about a fuller sample of objects; then the parameter value characterizing the operator can be optimized by a method of the "least-squares" kind.

If the decision maker is coherent in his choice, we can then classify the elements of Ω by order of preference, which should correspond to what the decision maker would have done spontaneously, had Ω contained few objects.

3.1.5. Example

Consider the problem of choosing a car on the basis of a catalogue of precise data concerning purchase price, fuel consumption, maximum speed, and responsiveness (evaluated as the time to reach 100 km/h from a standing start) for the available models.

We suppose that the desired characteristics (global objective) of the vehicle being sought are expressed in a simplified verbal form as

$$\langle \text{global objective} \rangle = (\langle \text{objective} \rangle \,|\, (\langle \text{objective} \rangle \langle \text{op} \rangle \langle \text{objective} \rangle))$$

where

$$\langle \text{objective} \rangle = (\langle \text{elementary objective} \rangle \,|\, (\langle \text{objective} \rangle \langle \text{op} \rangle \langle \text{objective} \rangle))$$

and "|" means "or." An elementary objective is the name of a fuzzy set on one of the four scales (price, consumption, speed, responsiveness); $\langle op \rangle$ is a verbal aggregation operator. Thus a global objective can be represented as a binary tree whose leaves (terminal nodes) are the elementary objectives, and each branch-point corresponds to an operation of aggregation. An example of such a global objective is

((fairly fast and responsive) but not too expensive)

Each elementary objective is identified by a dialogue with the decision maker, which yields a range of preferred values (equivalent in his own eyes), together with the set of definitely unacceptable values, which respectively determine the core and the support of the elementary fuzzy set. A further dialogue then serves to refine the form of the membership function.

Selection of an aggregation operator can be carried out by the following procedure. Three typical vehicles V_1, V_2, V_3 are presented to the decision maker for him to evaluate in terms of combined objectives G_1 and G_2 linked by means of the aggregation operator to be identified. The evaluation of each typical vehicle is an element of the set of levels A, B, C, D, E of Table 3.1 (see Section 3.1.1). The typical vehicles are chosen so as to enable discrimination between aggregation operators in a given list. The compatibility of each typical vehicle with each of the objectives to be combined is supposed to be known. In particular, the vehicles are supposed to be chosen so that

- V_1 is incompatible (score E) with G_1 but completely compatible (score A) with G_2.
- V_2 has medium compatibility (score C) with each of G_1 and G_2.
- V_3 has medium compatibility (score C) with G_1 and is completely compatible (score A) with G_2.

The aggregation operator h between μ_{G_1} and μ_{G_2} is then approximated by a function $\hat{h}$ from $\{A, C, E\}^2$ into $\{A, B, C, D, E\}$, which satisfies Axioms A1, A2, and A3 of Section 1.1. the decision maker thus provides the three values $\hat{h}(E, A)$, $\hat{h}(C, C)$, and $\hat{h}(C, A)$. Each triplet of replies corresponds to a standard aggregation operator as indicated in Table 3.2. In an appendix to this chapter we give a BASIC program which performs the identification of aggregation operators according to this principle, handles the queries relative to choice of vehicle, and then proposes a final ranking. This program corresponds to the first stage of identification (cf. Section 3.1.4).

There is a certain analogy between this approach and that of

TABLE 3.2. Selection of Aggregation Operations

	Vehicle type			
	V_1	V_2	V_3	
Objective G_1	E	C	C	Selected operators
Objective G_2	A	C	A	
Examples of	E	E	C	$\max(0, u + v - 1)$
possible responses	E	D	C	$u \cdot v$
by the decision	E	C	C	$\min(u, v)$
maker	E	C	B	$(u \cdot v)^{1/2}, 2uv/(u + v)$
	D	C	C	$\text{med}(u, v, 1/4)$
	C	C	C	$\text{med}(u, v, 1/2), \quad \min(u, v)/(1 - |u - v|)$
	C	C	B	$\frac{1}{2}(u + v), \sigma_{\div}, \max(u, v)/(1 + |u - v|)$
	C	C	A	σ_0
	B	C	B	$\text{med}(u, v, 3/4)$
	A	C	A	$\max(u, v), 1 - [(1 - u)(1 - v)]^{1/2}$
	A	B	A	$u + v - uv$
	A	A	A	$\min(1, u + v)$

"Karnaugh Tables" as used in the synthesis of logical circuits. We are here constructing a filter for a multivalued rather than a binary logic. The function that the filter realizes is supposed to represent the behavior of the decision maker when faced with the various objects that the computer can present to him. The class of available operations is viewed as a collection of "standard functions" like those that would be provided by a catalogue of logical circuits.

Table 3.2 is far from exhaustive and covers only some of the possible responses by the decision maker. The full list of possible responses contains 50 triples corresponding to the following constraints: (1) $\hat{h}(C, A) \geqslant \max(\hat{h}(E, A), \hat{h}(C, C))$; (2) $\hat{h}$ is symmetric; and (3) $\hat{h}(C, A) \geqslant C$ (meeting objective G_2 completely cannot depress the global level of satisfaction below the level of satisfaction of objective G_1). Note also that the function $\hat{h}$ is not completely defined if only three typical vehicles are used. Complete specification of $\hat{h}$ requires, in view of Axioms A1, A2, and A3, knowledge of the value of $\hat{h}(E, C)$ as well as the three values provided by the decision maker. The extra information would enable a finer discrimination to be made, but would raise the number of possible responses to 93.

The four classes of operator, namely, conjunctions, disjunctions, averages, and symmetric sums, only cover some of the 50 possible triples. Nevertheless, many of these triples correspond to minor modifications of standard operators [e.g., (D, C, C) is very close to (E, C, C), which is represented by the min operator]. But certain triples, fortunately few, such as (C, E, C), fall outside these four classes. An axiomatic approach to operators that correspond to such cases is yet to be developed.

It appears to be established that a very precise identification of an aggregation operator, corresponding to the second stage of the process (cf. Section 1.4), may sometimes (and perhaps often) be impossible. For all that, the approach is not invalidated, for two reasons:

- A very precise identification may be considered of little use; we only need to realize certain properties of an operator. This has to do with the arbitrary nature of the values of the membership function, and with the numerical representation of subjective judgments.
- If it is found that several different operators give possible models, it can be admitted that there is imprecision in the result of the aggregation. For example, if we have been unable to discriminate between operators $h_1, \ldots, h_k$, we could put

$$\forall (u, v), \qquad h(u, v) \in \left[\min_{i=1,k} h_i(u, v), \max_{i=1,k} h_i(u, v) \right]$$

A possible result of aggregation could also be a fuzzy number. This would be so if, among the selected operators, some were considered more credible than others. Thus we would end up with a fuzzy subset of operators which, by use of the extension principle, would give fuzzy evaluations of the objects.

The choice of objectives (i.e., determination of the μ_{G_i}) interacts with the choice of aggregation operator: a conjunction of undemanding objectives can be the same as a compromise between very selective objectives. This intuitive feature appears clearly in the model. For example, there is equivalence between $(\mu_{G_1}, \mu_{G_2}, h = \text{geometric mean})$ and $(\sqrt{\mu_{G_1}}, \sqrt{\mu_{G_2}}, h = \text{product})$. There would be some interest in a systematic search for such equivalences. In any case, it is clear that the way one combines objectives depends on how they have been specified.

3.2. Comparison of Imprecise Evaluations

In practice, it is rare to obtain a precise number for the result of a global evaluation of an object by a process of aggregation of criteria, or of uncertain factors. In general, the result will have a natural representation as a fuzzy interval expressing the imprecision and/or the uncertainty of the knowledge-base underlying the evaluation process, owing to inaccuracy of measurement, verbal data, incomplete definition of mode of aggregation, or uncertainty about the expected properties of the objects.

The aggregation procedures discussed in the first part of this chapter are still valid when the objects have imprecise partial evaluations, which can be represented by fuzzy quantities on the reference set X_i associated with the partial objective G_i. The global evaluation is then obtained by applying the extension principle (cf. Chapter 2) to the aggregation operator $h(\mu_{G_1}(\cdot), \ldots, \mu_{G_q}(\cdot))$. Let $F_1, \ldots, F_q$ be the fuzzy evaluations of the object ω in terms of the q criteria. The *degree of compatibility* of the objective G_i with the fuzzy evaluation $m_i(\omega) = F_i$ is $\tau_i = \mu_{G_i}(F_i)$ defined by

$$\forall t \in [0, 1], \qquad \mu_{\tau_i}(t) = \begin{cases} \sup\{\mu_{F_i}(u) \mid t = \mu_{G_i}(u)\} \\ 0 \qquad \text{if } \nexists u, t = \mu_{G_i}(u) \end{cases} \qquad (3.19)$$

This quantity will turn up again in Chapter 4. The global evaluation of the object ω will then be

$$\mu_D(\omega) = h(\tau_1, \ldots, \tau_q) \qquad (3.20)$$

where $\mu_D(\omega)$ is a fuzzy quantity on $[0, 1]$ which is obtained by applying the results of the calculus of fuzzy quantities (cf. Chapter 2) to the isotonic function h. For example, if h is the min operator, we get

$$\mu_D(\omega) = \widetilde{\min}\, (\tau_1, \ldots, \tau_q)$$

When the global evaluations are imprecise, ranking the objects in terms of these evaluations is not trivial, since the set of fuzzy quantities has no natural total-ordering structure. Our exposé is based on the results of Dubois and Prade [9], which brings together and completes the approaches of Baas and Kwakernaak [2], Tsukamoto, Nikiforuk, and Gupta [28] and Watson, Donnell, and Weiss [30]. An extensive survey of comparison techniques has been recently published by Bortolan and Degani [42].

3.2.1. Comparison of a Real Number and a Fuzzy Interval

Before we can compare two fuzzy intervals, it is first appropriate to know how a real number may be placed in relation to a fuzzy interval. Possibility calculus enables us to define the set of numbers that *can* be greater than or equal to the values of a variable associated with a fuzzy interval P, and the set of numbers that are *necessarily* greater. Denote these, respectively by $[P, +\infty), \,]P, +\infty)$; they are such that [9, 10]

$$\forall \, w, \qquad \mu_{[P, +\infty)}(w) = \Pi_P((-\infty, w]) = \sup_{u \leqslant w} \mu_P(u) \qquad (3.21)$$

$$\mu_{]P, +\infty)}(w) = N_P((-\infty, w[) = \inf_{u \geqslant w} (1 - \mu_P(u)) \qquad (3.22)$$

where Π_P and N_P are the possibility and necessity measures defined in terms of the distribution μ_P. $[P, +\infty)$ and $]P, +\infty)$ are convex, and their membership functions are the upper and lower distribution functions of P (cf. Section 1.1 of Chapter 2), except possibly at points of discontinuity: see Figure 3.14.

Similarly, we can define $(-\infty, P]$ as the fuzzy set of numbers possibly, and $(-\infty, P[$ as the fuzzy set of numbers necessarily less than P, which are such that

$$(-\infty, P] = \overline{]P, +\infty)} \qquad (3.23)$$

$$(-\infty, P[= \overline{[P, +\infty)} \qquad (3.24)$$

where the bar denotes fuzzy set-theoretic complementation.

This symbolism is justified by the fact that if P reduces to a real number u, then $[P, +\infty)$ reduces to the half-line $[u, +\infty)$, and $]P, +\infty)$ to the half-line $]u, +\infty)$. It can be verified that

$$(-\infty, P] \cap [P, +\infty) = P$$

$$(-\infty, P[\, \cap \,]P, +\infty) = \varnothing$$

where the intersection is taken in the min sense.

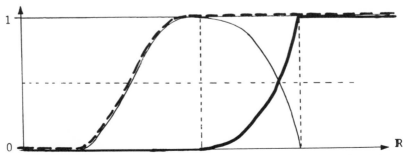

Figure 3.14. Fuzzy intervals derived from a fuzzy bound. ——, P; – – –, $[P, +\infty)$; ——, $]P, +\infty)$.

When P is fuzzy, $P \cup]P, +\infty) \subset [P, +\infty)$ in general (strict inclusion, unless $\cup$ is defined, for example, by the bounded sum).

We can even define the domain (possibly empty) bounded by two fuzzy intervals P and Q. Taking P as the lower and Q as the upper bound of the domain, the fuzzy "closure" and "interior" of the domain are, respectively, $[P, Q]$ and $]P, Q[$ defined (cf. [10]) by

$$[P, Q] = (-\infty, Q] \cap [P, +\infty)$$
$$]P, Q[= (-\infty, Q[\cap]P, +\infty)$$

where the intersection is defined in terms of min.

3.2.2. Comparison of Two Fuzzy Intervals

To the extent that they may overlap each other, there is not necessarily a clear answer to the question of finding the larger of two fuzzy intervals. A natural first idea is to use $\widetilde{\max}$ and $\widetilde{\min}$ (see Freeling [13]). However, it can easily be seen that two closely similar fuzzy intervals, which overlap substantially, can after all be discriminated by $\widetilde{\max}$ and $\widetilde{\min}$. For example, $\widetilde{\max}(P, Q) = Q$ when P is a fuzzy interval and also $\exists \varepsilon > 0: \mu_Q(w) = \mu_P(w - \varepsilon), \forall w$. On the other hand, $\widetilde{\max}$ no longer discriminates when the ascending or descending arms of the membership functions intersect, however far separated the cores of P and Q may be. What is really needed is a basis for a possible quantitative discrimination, which cannot be made with $\widetilde{\max}$ and $\widetilde{\min}$.

In terms of possibility theory, to determine whether P is greater than Q one can compare on the one hand the sets P and $[Q, +\infty)$, and on the other P and $]Q, +\infty)$, by means of such an index of comparison as the possibility or necessity of a fuzzy event (cf. Chapter 1, Section 1.7), i.e., calculate the possibility and the necessity, in the sense of μ_P, of the fuzzy events $[Q, +\infty)$ and $]Q, +\infty)$. In this way we obtain the four fundamental indices of comparison:

1. $\Pi_P([Q, +\infty)) = \sup_u \min\left(\mu_P(u), \sup_{v \leq u} \mu_Q(v) \right)$

$$= \sup_{u \geq v} \min(\mu_P(u), \mu_Q(v)) \qquad (3.25)$$

If X and Y are variables whose domains are constrained by μ_P and μ_Q, respectively, then $\Pi_P([Q, +\infty))$ is interpreted as the possibility $\mathrm{Pos}(\bar{X} \geq \bar{Y})$ that X should take a value at least as large as Y, i.e., that the largest values that X can take are at least as great as the smallest values that Y can take. If μ_P and μ_Q are continuous functions, then $\text{Pos}(\bar{X} \geq \underline{Y}) = \text{Pos}(\bar{X} > \underline{Y})$, where $\text{Pos}(\bar{X} > \underline{Y})$ is defined by changing "$u \geq v$" into "$u > v$" in the definition of $\Pi_P([Q, +\infty))$.

2. $\Pi_P(]Q, +\infty)) = \sup_u \min\left(\mu_P(u), \inf_{v \geq u}[1 - \mu_Q(v)]\right)$

$$= \sup_u \inf_{v \geq u} \min(\mu_P(u), 1 - \mu_Q(v)) \qquad (3.26)$$

This quantity can be interpreted as the possibility that the largest values that X can take are greater than the largest values that Y can take (see Figure 3.15); this possibility can be written symbolically as $\text{Pos}(\bar{X} > \bar{Y})$.

3. $N_P([Q, +\infty)) = \inf_u \max\left(1 - \mu_P(u), \sup_{v \leq u} \mu_Q(v)\right)$

$$= \inf_u \sup_{v \leq u} \max(1 - \mu_P(u), \mu_Q(v)) \qquad (3.27)$$

This quantity is interpreted as the necessity that the smallest values that X can take are at least equal to the smallest values that Y can take (see Figure 3.15); this necessity is written $\text{Nec}(\underline{X} \geq \underline{Y})$.

4. $N_P(]Q, +\infty)) = \inf_u \max\left(1 - \mu_P(u), \inf_{v \geq u}[1 - \mu_Q(v)]\right)$

$$= 1 - \sup_{u \leq v} \min(\mu_P(u), \mu_Q(v)) \qquad (3.28)$$

This quantity is interpreted as the necessity $\text{Nec}(\underline{X} > \bar{Y})$ that X can take

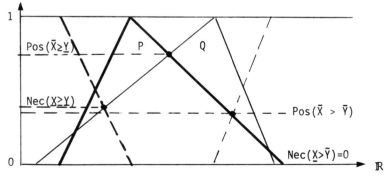

Figure 3.15. Comparison of two fuzzy numbers.

only values that are greater than values that Y can take, i.e., the necessity that the smallest values X can take are greater than the largest that Y can take. If μ_P and μ_Q are continuous functions, then $\text{Nec}(\underline{X} > \bar{Y}) = \text{Nec}(\underline{X} \geq \bar{Y})$, where $\text{Nec}(\underline{X} \geq \bar{Y})$ is defined by changing "$u \leq v$" into "$u < v$" in the expression for $N_P(]Q, +\infty))$.

In fact the possibilistic approach enables 16 possible values to be constructed which describe the relative positions of P and Q. To generate them it suffices to change P into Q or $[P, +\infty)$ into $(-\infty, P]$ etc., in (3.25)–(3.28).

A study of these 16 possibilities (Dubois and Prade [9]) shows that only six of them are independent, in general, as Table 3.3 suggests (where P and Q play symmetric rôles). These six values are necessary and sufficient to characterize the relative positions of two arbitrary nonfuzzy intervals. For fuzzy intervals, however, most of the time we have

$$\Pi_P(]Q, +\infty)) = 1 - \Pi_Q(]P, +\infty)) \tag{3.29}$$

$$N_P([Q, +\infty)) = 1 - N_Q([P, +\infty)) \tag{3.30}$$

The two indices appearing in the first equation above enable us, when equality does not hold, to distinguish the relative positions of intervals of the form $P = [a_1, b]$ and $Q = [a_2, b]$ [where $\Pi_P(]Q, +\infty)] = 0 = \Pi_Q(]P, +\infty))$, and of intervals of the form $P = [a_1, b[$ and $Q = [a_2, b]$ [where $\Pi_Q(]P, +\infty)) = 1$].

A similar remark holds for the indices in the second equation. In practice, the four indices $\text{Pos}(\bar{X} \geq \underline{Y})$, $\text{Pos}(\bar{X} > \bar{Y})$, $\text{Nec}(\underline{X} \geq \underline{Y})$, and $\text{Nec}(\underline{X} > \bar{Y})$ suffice for discussing the relative positions of two fuzzy intervals. These indices have the following properties:

- $\text{Pos}(\bar{X} \geq \underline{Y}) \geq \max(\text{Pos}(\bar{X} > \bar{Y}), \text{Nec}(\underline{X} \geq \underline{Y}))$;
- $\min(\text{Pos}(\bar{X} > \bar{Y}), \text{Nec}(\underline{X} \geq \underline{Y})) \geq \text{Nec}(\underline{X} > \bar{Y})$;
- $\text{Pos}(\bar{X} \geq \underline{Y}) = 1 - \text{Nec}(\underline{Y} > \bar{X})$;
- $\max(\text{Pos}(\bar{X} \geq \underline{Y}), \text{Pos}(\bar{Y} \geq \underline{X})) = 1$;
- $\text{Pos}(\bar{X} > \bar{Y}) + \text{Pos}(\bar{Y} > \bar{X}) = 1$;
- $\text{Nec}(\underline{X} \geq \underline{Y}) + \text{Nec}(\underline{Y} \geq \underline{X}) = 1$.

TABLE 3.3. Properties of Ranking Indices

$\Pi_P([Q, +\infty)) = 1 - N_P((-\infty, Q[) = 1 - N_Q(]P, +\infty)) = \Pi_Q((-\infty, P])$
$\Pi_P(]Q, +\infty)) = 1 - N_P((-\infty, Q]) \leq 1 - \Pi_Q(]P, +\infty)) = N_Q((-\infty, P])$
$N_P([Q, +\infty)) = 1 - \Pi_P((-\infty, Q[) \geq 1 - N_Q([P, +\infty)) = \Pi_Q((-\infty, P[)$
$N_P(]Q, +\infty)) = 1 - \Pi_P((-\infty, Q]) = 1 - \Pi_Q([P, +\infty)) = N_Q((-\infty, P[)$

The first four simply reexpress characteristic properties of possibility and necessity measures in the context of comparison of fuzzy intervals. The last two simply rewrite (3.29) and (3.30) and hold under the same conditions.

We can similarly introduce indices of equality between the fuzzy intervals P and Q. Three indices can be naturally envisaged:

1. $\Pi_Q(P) = \Pi_P(Q)$, i.e., the possibility of the fuzzy event P under the distribution μ_Q, and the possibility of Q under μ_P. This quantity can also be written $\text{Pos}(X = Y)$ in terms of the variables X and Y constrained by μ_P and μ_Q, respectively, i.e., the possibility of X and Y both taking at least one common value.
2. $N_Q(P)$, i.e., the necessity of the fuzzy event P under the distribution μ_Q. This is an index of inclusion of Q in P, interpreted as the degree of certainty that, given a value taken by Y, the same value can be taken by X.
3. $N_P(Q)$, an index of inclusion of P in Q, with a similar interpretation.

The lack of symmetry of the last two naturally leads to constructing a symmetric combination of them, of conjunctive type:

$$\text{Nec}(X = Y) = \min(N_P(Q), N_Q(P))$$

The use of the min conjunction is justified by the fact that it is idempotent; when $P = Q$, i.e., $\mu_P = \mu_Q$, we can expect the equality $\text{Nec}(X = Y) = N_Q(Q)$. $\text{Nec}(X = Y)$ measures the degree of certainty that X and Y may take a common value compatible with P and Q, whatever the value may be.

These indices of equality are easily recovered from the indices of inequality introduced earlier, since we can verify that

$$\text{Pos}(X = Y) = \min(\text{Pos}(\bar{X} \geq \underline{Y}), 1 - \text{Nec}(\underline{X} > \bar{Y}))$$
$$N_Q(P) = \min(\text{Nec}(\underline{Y} \geq \underline{X}), 1 - \text{Pos}(\bar{Y} > \bar{X}))$$
$$\text{Nec}(X = Y) = \min(\text{Nec}(\underline{X} \geq \underline{Y}), 1 - \text{Pos}(\bar{X} > \bar{Y}),$$
$$1 - \text{Pos}(\bar{Y} > \bar{X}), \text{Nec}(\underline{Y} \geq \underline{X}))$$

Thus once again the four indices of inequality contain the necessary information for a discussion of equality between fuzzy intervals.

An experimental study of various techniques for comparing fuzzy intervals is reported in [42], where the approach based on the four indices proposed here never produces counterintuitive rankings with normalized fuzzy intervals.

N.B. An index of comparison between fuzzy quantities may acquire gradation if the relation "$\geq$" in (3.25)–(3.28) is replaced by a fuzzy relation expressing to what extent it is desired that one fuzzy quantity should exceed another. Another way of proceeding involves modifying P, Q, $[P, +\infty)$, $]P, +\infty)$, etc. by means of a fuzzy relation of proximity such as the one considered in Section 2.3.2 of Chapter 2, whose effect would be to make the possibility distributions attached to these fuzzy sets less restrictive. The latter idea was suggested by Baldwin and Guild [3]. □

3.2.3. Ordering of n Fuzzy Intervals

Two approaches can be envisaged for ordering a set of n fuzzy intervals $\{M_1, \ldots, M_n\}$ by taking advantage of the indices that have been introduced:

- Define global indices of exceedance that evaluate the degree to which M_i dominates all the other fuzzy intervals.
- Use the fuzzy relations obtained by comparing the M_i two by two.

Only the first approach will be considered here; the second has been dealt with in detail in Dubois and Prade [9] and Roubens and Vincke [43].

An index of global exceedance can be quite naturally defined by interpreting the domination by M_i of all the other fuzzy intervals as the fact that M_i exceeds "the largest of the M_j," $j \neq i$. The $\widetilde{\max}$ operator naturally intervenes here: four indices of exceedance can be constructed from (3.25)–(3.28) by comparing M_i and $\widetilde{\max} M_j$ ($j \neq i$):

$$\text{PSE}(M_i) = \Pi_{M_i}\left(\left[\widetilde{\max_{j \neq i}} M_j, +\infty\right)\right) \qquad \text{(exceedance possibility)}$$

$$\text{PS}(M_i) = \Pi_{M_i}\left(\left]\widetilde{\max_{j \neq i}} M_j, +\infty\right)\right) \qquad \text{(strict exceedance possibility)}$$

$$\text{NSE}(M_i) = N_{M_i}\left(\left[\widetilde{\max_{j \neq i}} M_j, +\infty\right)\right) \qquad \text{(exceedance necessity)}$$

$$\text{NS}(M_i) = N_{M_i}\left(\left]\widetilde{\max_{j \neq i}} M_j, +\infty\right)\right) \qquad \text{(strict exceedance necessity)}$$

The set $\{M_1, \ldots, M_n\}$ can be ordered according to the values of each index. Thus we obtain four linear orderings which, if they are coherent, give rise to a definite ordering. Nonetheless, owing to imprecision, these four indices will not necessarily give the same ordering (see the example below). Observe that the expression for the first index simplifies to

$$\text{PSE}(M_i) = \min_{j \neq i} \Pi_{M_i}([M_j, +\infty))$$

3.2.4. Computer Implementation

Explicit calculation of the values of the four indices reduces to finding the coordinates of the points of intersection of membership functions. When the fuzzy intervals are of type LR, i.e., $P = (p, \bar{p}, \alpha, \beta)_{LR}$, $Q = (q, \bar{q}, \gamma, \delta)_{LR}$, we have to solve the following equations:

- Find u_1 such that

$$R\left(\frac{u_1 - \bar{p}}{\beta}\right) = L\left(\frac{q - u_1}{\gamma}\right) \qquad [= \text{Pos}(\bar{X} \geqslant Y)]$$

  (intersection between right-hand arm of μ_P and left-hand arm of μ_Q).
- Find u_2 such that

$$L\left(\frac{p - u_2}{\alpha}\right) = 1 - L\left(\frac{q - u_2}{\gamma}\right) \qquad [= \text{Nec}(X \geqslant Y)]$$

  (intersection between the left-hand arms of μ_P and μ_Q).
- Find u_3 such that

$$1 - R\left(\frac{u_3 - \bar{p}}{\beta}\right) = R\left(\frac{u_3 - \bar{q}}{\delta}\right) \qquad [= \text{Pos}(\bar{X} > \bar{Y})]$$

  (intersection between the right-hand arms of μ_P and μ_Q).
- Find u_4 such that

$$L\left(\frac{p - u_4}{\alpha}\right) = R\left(\frac{u_4 - \bar{q}}{\delta}\right) \qquad [= \text{Nec}(X > \bar{Y})]$$

  (intersection between the left-hand arm of μ_P and the right-hand arm of μ_Q).

The BASIC program which appears in the appendix to this chapter calculates these indices for trapezoidal fuzzy intervals $[L(u) = R(u) = \max(0, 1 - u)]$. It is easy to obtain the following expressions for the indices in terms of the spreads and modal values of the fuzzy intervals:

$$\text{Pos}(\bar{X} \geqslant \underline{Y}) = \max\left(0, \min\left(1, 1 + \frac{(\bar{p} - \underline{q})}{(\beta + \gamma)}\right)\right) \quad \text{(PSE)}$$

$$\text{Nes}(\underline{X} \geqslant \underline{Y}) = \max\left(0, \min\left(1, \frac{(\underline{p} - \underline{q} + \gamma)}{(\alpha + \gamma)}\right)\right) \quad \text{(NSE)}$$

$$\text{Pos}(\bar{X} > \bar{Y}) = \max\left(0, \min\left(1, \frac{(\bar{p} - \bar{q} + \beta)}{(\beta + \delta)}\right)\right) \quad \text{(PS)}$$

$$\text{Nes}(\underline{X} > \bar{Y}) = \max\left(0, \min\left(1, \frac{(\underline{p} - \bar{q})}{(\alpha + \delta)}\right)\right) \quad \text{(NS)}$$

These formulas hold except when the sums of spreads in the denominators are zero, which occurs when P and Q are ordinary closed intervals. If one of the denominators is zero, fictitious spreads are calculated, which are used in place of the zero spreads, so that the desired points of intersection exist. In order not to perturb the results, these points of intersection (theoretically at infinity) should have coordinates outside the interval $[0, 1]$. For this, the fictitious spreads should have a value ε such that

$$\varepsilon < \varepsilon_0 = \tfrac{1}{2} \min(|\underline{p} - \underline{q}|, |\underline{p} - \bar{q}|, |\bar{p} - \underline{q}|, |\bar{p} - \bar{q}|)$$

If $\varepsilon_0 = 0$, then

- If $\bar{p} = \underline{q}$, $\beta + \gamma = 0$, we can substitute any positive values for β and γ to calculate $\text{Pos}(\bar{X} \geqslant \underline{Y})$.
- If $\underline{p} = \bar{q}$, $\alpha + \delta = 0$, we can substitute any positive values for α and δ to calculate $\text{Nec}(\underline{X} > \bar{Y})$.
- If $\underline{p} = \underline{q}$, $\alpha + \gamma = 0$, then $\text{Nec}(\underline{X} \geqslant \underline{Y}) = 0$ as can be verified from the theoretical formulas.
- If $\bar{p} = \bar{q}$, $\beta + \delta = 0$, then $\text{Pos}(\bar{X} > \bar{Y}) = 0$ as can be verified from the theoretical formulas.

3.2.5. Example

Consider the three fuzzy intervals N_1, N_2, and N_3 of Figure 3.16, which restrict the domains of the three variables X_1, X_2, and X_3 respectively. We wish to find how much bigger each interval is than the others. The values of the four indices are calculated to be as shown in Table 3.4. Because the intervals overlap each other, none strictly dominates the others in respect of necessity ($\text{NS} = 0$). Ordering by the index PS clearly shows that the largest values that X_3 can take are

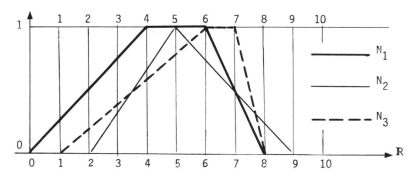

Figure 3.16. Example.

TABLE 3.4. Results of the Example

	PSE	PS	NSE	NS
N_1	1	1/3	2/7	0
N_2	8/9	2/5	1/2	0
N_3	1	3/5	1/2	0

generally greater than those the other variables can take. Ordering by the index NSE sets N_1 aside, and ordering by PSE sets N_2 aside. Thus we find, as a glance at Figure 3.16 suggests, that N_3 can be considered as the largest of these fuzzy intervals.

Appendix: Computer Programs

1.

```
REM ================================================================
REM                         GOAL AGGREGATION
REM ================================================================
STRSIZ 80

REM ----------------------------------------------------------------
REM                 Definition of arrays encoding the tree
REM ----------------------------------------------------------------
DIM PERE%(15)                       ! Pointers to ancestors
DIM CRIT$(15)                       ! Literal chain denoting a goal
DIM CRIT%(15)                       ! Index denoting a criterion or an operation
DIM ACCEP(15,4)                     ! Array which stores the values which express
                                    ! the level of satisfaction of simple goal
DIM ECHELLE(4,2)                    ! Array of measurement scales for goals


REM Constants :
REM -----------
DIM UNITE$(4)
UNITE$(1)=" F "                     ! Price unit
ECHELLE(1,1)=0:ECHELLE(1,2)=500000
UNITE$(2)=" sec aux 100 Km/h"       ! Responsiveness unit (acceleration)
ECHELLE(2,1)=0:ECHELLE(2,2)=100
```

```
UNITE$(3)=" km/h"                    ! Speed unit
ECHELLE(3,1)=0:ECHELLE(3,2)=350
UNITE$(4)=" litres aux 100km "       ! Consumption unit
ECHELLE(4,1)=0:ECHELLE(4,2)=30
NBOP=12                              ! Number of operations

REM   Acquisition of response triples for each operation
REM   -----------------------------------------------------
DIM OP%(NBOP,3)                      ! Array of interpretable response  triples
FOR J=1 TO NBOP
  READ OP%(J,1):READ OP%(J,2):READ OP%(J,3)
NEXT J

REM Acquisition of characteristics for each alternative in the database
REM -----------------------------------------------------------------------
DIM NOM$(15)                         ! Car type
DIM PERF(15,4)                       ! Table of characteristics of each type
DIM FILE(15,6)                       ! Stacks for computing global evaluation for
                                     ! each car type.

FOR J=1 TO 15
  READ NOM$(J)
  READ PERF(J,1)                     ! Read price
  READ PERF(J,2)                     ! Read acceleration
  READ PERF(J,3)                     ! Read speed
  READ PERF(J,4)                     ! Read consumption
NEXT J


REM ==========================================================================
REM                     QUERY INTERPRETATION PROGRAM
REM ==========================================================================
PRINT TAB(-1,0)                      ! Clear the screen
PRINT TAB(1,15);"DECISION-SUPPORT FOR CHOOSING A CAR"
PRINT TAB(2,15);"========================================"
PRINT TAB(4,1); "Input goals :"
INPUT R$                             ! Query input
N=0                                  ! Initialization of current node variable
NO=0                                 ! Initialization of current father node
I1=0                                 ! Word pointer in the query chain
SOMMET PILE=0

REM ------------------------------------------------------------------------
REM   Patterns of the form (<crit>op<crit>) are searched for. Four kinds of
REM   processing are considered according to the nature of the word which is
REM   read
REM   - an opening parenthesis    - a linguistic operation
REM   - a closing parenthesis     - a simple goal.
REM
REM ------------------------------------------------------------------------

REM   Read next left-most word :
REM   --------------------------
                                     ! words are separated by blanks
PRMOT:
  I=I1+1
  I1=INSTR(I,R$," ")
  MOTLU$=R$[I,I1-1]                  ! MOTLU$ is the  leftmost  word read.

REM   Case when the word  is an opening parenthesis :
REM   -----------------------------------------------
PAROUV:
  IF MOTLU$<>"(" THEN GOTO PARFER
    N=N+1                            ! Creation of a new node
    PERE%(N)=NO                      ! Updating of father pointer
    CRIT$(N)="("                     ! The chain starts being filled
    NO=N                             ! Down into the tree
    MOTPREC$="("                     ! Updating of last word and
    GOTO PRMOT                       ! reading of the next one

REM Case of a closing parenthesis :
REM -------------------------------
PARFER:
  IF MOTLU$<>")" THEN GOTO MOTORD
    CRIT$(NO)=CRIT$(NO)+" "+")"      ! Completion of the word chain
```

```
    IF NO<>1 THEN CRIT$(PERE%(NO))=CRIT$(PERE%(NO))+" "+CRIT$(NO)
        ! if the node is not the root, completion of the father node chain

    GOSUB AGREG                             ! Goal aggregation is performed.
        ! Once the operation is identified, each car is rated according to
        ! this compound goal in respective stacks
    GOSUB EVALRESAGR
    NO=PERE%(NO)                            ! Back up in the tree
    MOTPREC$=")"                            ! Updating of preceding word
    IF NO<>0 THEN GOTO PRMOT                ! If it is not the end, read next word
                                            ! Otherwise the final ranking is per-
    GOTO CLASSEMENT                         ! formed.

REM Case of a regular word :
REM --------------------- -
MOTORD:
    IF (MOTPREC$<>"CRITERE" AND MOTPREC$<>")") THEN GOTO CRIT
        ! Otherwise the word read is an aggregation operation
        CRIT$(NO)=CRIT$(NO)+" "+MOTLU$+" " ! updating of the chain
        MOTPREC$="OPERATEUR"               ! updating of preceding word
        GOTO PRMOT                         ! next word is read
REM Processing of a simple goal
REM ----------------------------
CRIT:
    ! It is a new goal
    N=N+1                                   ! A new node is created
    PERE%(N)=NO                             ! Update father pointer
    REP'MOT'CLEF:
        ! Detection of a key word:
        IF MOTLU$= "DEAR"   THEN CRIT%(N)=1:CRIT$(N)=CRIT$(N)+" "+MOTLU$:GOTO FIN
        IF MOTLU$= "SWIFT"   THEN CRIT%(N)=2:CRIT$(N)=CRIT$(N)+" "+MOTLU$:GOTO FIN
        IF MOTLU$= "FAST"   THEN CRIT%(N)=3:CRIT$(N)=CRIT$(N)+" "+MOTLU$:GOTO FIN
        IF MOTLU$= "ECONOMICAL" THEN CRIT%(N)=4:CRIT$(N)=CRIT$(N)+" "+MOTLU$:GOTO FIN
        ! If the word is not a key word, it is stored in the chain and the next one
        ! is considered
        CRIT$(N)=CRIT$(N)+" "+MOTLU$
        I=I1+1
        I1=INSTR(I,R$," ")
        MOTLU$=R$[I,I1-1]
        GOTO REP'MOT'CLEF

    FIN:                                    ! The goal is now identified
        CRIT$(NO)=CRIT$(NO)+CRIT$(N)        ! The father chain is updated
        MOTPREC$="CRITERE"                  ! Updating of preceding word
        GOSUB PRECIS                        ! An acquisition routine is run for
                                            ! extra information
        SOMMET'PILE=SOMMET'PILE+1

REM Rating of each car for the simple goal :
REM -------------------------------------------
    FOR J=1 TO 15
    LET P = PERF(J,CRIT%(N))
    IF (P<=ACCEP(N,3) OR P >=ACCEP(N,4)) THEN PERF=0
    IF (P>=ACCEP(N,1) AND P<=ACCEP(N,2)) THEN PERF=1
    IF (P>ACCEP(N,3) AND P<ACCEP(N,1)) THEN PERF=(P-ACCEP(N,3))/(ACCEP(N,1)-ACCEP(N,3
    IF (P>ACCEP(N,2) AND P<ACCEP(N,4)) THEN PERF=(ACCEP(N,4)-P)/(ACCEP(N,4)-ACCEP(N,2
    LET PILE(J,SOMMET'PILE)=PERF
    NEXT J

        GOTO PRMOT


CLASSEMENT:
    PRINT TAB(-1,0);" RANKING OF THE 15 CURRENT CARS : "
    PRINT "-------------------------------------------
    FOR I=1 TO 15
        M=-1                                ! Initiating the maximal evaluation
        IM=0                                ! corresponding index
        FOR J=1 TO 15
            IF PILE(J,1)>M THEN M=PILE(J,1):IM=J
        NEXT J
        PRINT I;". ";NOM$(IM);
        PRINT TAB(25);PERF(IM,1);
```

```
        PRINT TAB(35);PERF(IM,2);
        PRINT TAB(45);PERF(IM,3);
        PRINT TAB(55);PERF(IM,4);
        PRINT TAB(65);
        PRINT USING "##.## %",100*PILE(IM,1)
        PILE(IM,1)=-1                        ! The value -1 is forced not to assign
                                             ! the same index twice.
    NEXT I
END


REM ----------------------------------------------------------------------
REM                        SUBROUTINE PRECIS
REM
REM   This subroutine enables more information to be obtained about the meaning of
REM   a simple goal. Especially it asks for the 4 values which characterize the
REM   trapezoid which describes the fuzzy goal.
REM
REM  Parameters : I, N
REM  Yields : ACCEP (N,1.,4) corresponding to the four values.
REM ----------------------------------------------------------------------
PRECIS:
    PRINT TAB(6,I+2);"^"
    PRINT TAB(7,1); "Indicate the concerned goal : (unit : ";&
      ECHELLE(CRIT%(N),1);" a ";ECHELLE(CRIT%(N),2); UNITE$(CRIT%(N));

    PRINT " )";TAB(-1,9)
    PRINT TAB(9,1); "Interval of maximal satisfaction : ";TAB(-1,9);
    INPUT ACCEP(N,1),ACCEP(N,2)
    PRINT TAB(10,1); "maximal interval allowed" : ";TAB(-1,9);
    INPUT ACCEP(N,3),ACCEP(N,4)
    !If the interval of maximal satisfaction touches one of the scale limits, the
    !spread on this side is modified.

    LGRECH=ECHELLE(CRIT%(N),2)-ECHELLE(CRIT%(N),1)
    IF ACCEP(N,1)=ECHELLE(CRIT%(N),1) THEN ACCEP(N,3)=ACCEP(N,3)-LGRECH
    IF ACCEP(N,2)=ECHELLE(CRIT%(N),2) THEN ACCEP(N,4)=ACCEP(N,4)+LGRECH
    PRINT TAB(6,I+2);" "
RETURN

REM ----------------------------------------------------------------------
REM                        SUBROUTINE AGREG
REM
REM  This subroutine determines, through a query, the index corresponding to the
REM  best-fitting operation.
REM

REM----------------------------------------------------------------------
AGREG:
    FILSG=NO+1                              ! determines the index of the left son
    FILSD=N                                 ! determine the index of the right son
    TEST'ASCENDANT:
       IF PEREX(FILSD)<>NO THEN FILSD=PEREX(FILSD) : GOTO TEST'ASCENDANT

    PRINT TAB(12,1); "AGGREGATION OF GOALS"
    PRINT "Objectif 1 : ";CRIT$(FILSG)
    PRINT "Objectif 2 : ";CRIT$(FILSD)
    PRINT "What is your rating of cars satisfying :"
    FOR I=1 TO 3
      PRINT TAB(17,10+20*I);"car";I;
      PRINT TAB(18,10+20*I);"-----";
    NEXT I
    PRINT TAB(19,1); "GOAL 1"
    PRINT "GOAL 2"
    I=1:F=FILSG:GOSUB EVALPERF
    I=2:F=FILSD:GOSUB EVALPERF
    PRINT TAB(24,1); "5 Very good / 4 Good / 3 Fairly good / 2 Poor / 1 Very bad
    REP:
      PRINT TAB(22,1); "YOUR RATING"
      PRINT TAB(22,30);
      INPUT R1
      PRINT TAB(22,50);
      INPUT R2
      PRINT TAB(22,70);
```

```
        INPUT R3
        I=1

        ! The response table is scanned. If a suitable operation is not found, new
        ! ratings are requested.
     A: IF OP%(I,1)<R1 THEN I=I+1:GOTO A
     B: IF OP%(I,2)<R2 THEN I=I+1:GOTO B
     C: IF OP%(I,3)<R3 THEN I=I+1:GOTO C
        IF OP%(I,1)<>R1 OR OP%(I,2)<>R2 OR OP%(I,3)<>R3 THEN &
             PRINT TAB(23,1); "THIS RESPONSE CANNOT BE INTERPRETED" : GOTO REP
   CRIT%(NO)=4+I
RETURN
REM --------------                -------------------------------------------------
REM                                SUBROUTINE EVALPERF
REM
REM    This subroutine calculates the ratings which characterize the prototypical
REM  cars. If the goal is simple, these are numerical values. If it is compound
REM  these are linguistic values. The exhibited prototypes depend on the ordering
REM  of the goals.
REM  Parameters : I = 1 or 2 whether it is the 1st or second goal
REM                CRIT %(I) : number of the goal if < 4, number of operation
REM                               otherwise.
REM                F index of the son node concerned.
REM  -----------------------------------------------------------------------------
EVALPERF:
    IF CRIT%(F)<=4 THEN GOTO CRITSPLE
        ! otherwise the goal is compound and the ratings are linguistic
    IF I=1 THEN PRINT TAB(19,30);TAB(-1,9); "VERY BAD": &
                   PRINT TAB(19,50);"FAIRLY GOOD" : PRINT TAB(19,70);"FAIRLY GOOD"; &
              ELSE PRINT TAB(20,30);TAB(-1,9); "VERY GOOD" : &
PRINT TAB(20,50);"FAIRLY GOOD":PRINT TAB(20,70);"VERY GOOD";
    RETURN

CRITSPLE:
    ! The case of a simple goal
    ALPHA=ACCEP(F,1)-ACCEP(F,3)
    BETA=ACCEP(F,4)-ACCEP(F,2)
    PLSTR=ALPHA MIN BETA                        ! the smallest spread is kept
    ! The three necessary values are established.
    IF PLSTR=ALPHA THEN V2=ACCEP(F,1)-ALPHA/2:V1=((ACCEP(F,1)-3*ALPHA MAX &
                             ECHELLE(CRIT%(F),1)) MIN ECHELLE(CRIT%(F),2)) &
                      ELSE V2=ACCEP(F,2)+BETA/2:V1=((ACCEP(F,2)+3*BETA MAX &
                             ECHELLE(CRIT%(F),1)) MIN ECHELLE(CRIT%(F),2))
       V3=(ACCEP(F,1)+ACCEP(F,2))/2
       IF I=2 THEN GOTO L2
                    PRINT TAB(19,30);V1;TAB(-1,9)
                    PRINT TAB(19,50);V2
                    PRINT TAB(19,70);V2
                    RETURN
            L2:  PRINT TAB(20,30);V3;TAB(-1,9)
                    PRINT TAB(20,50);V2
                    PRINT TAB(20,70);V3
    RETURN

REM  -----------------------------------------------------------------------------
REM                                SUBROUTINE EVAL/RESAGR
REM
REM    This subroutine performs the actual computation of the application  of the
REM  aggregation operation on the 2 operands located on top of each stack and
REM  puts the result on top of stack.
REM  -----------------------------------------------------------------------------
EVALRESAGR:
      FOR J=1 TO 15
         X=PILE(J,SOMMET'PILE)
         Y=PILE(J,SOMMET'PILE-1)
         IF CRIT%(NO)=5 THEN Y=0 MAX (Y+X-1)
         IF CRIT%(NO)=6 THEN Y=Y*X
         IF CRIT%(NO)=7 THEN Y=Y MIN X
         IF CRIT%(NO)=8 THEN Y=SQR(Y*X)
         IF CRIT%(NO)=9 THEN Y=Y+X+1/4-(Y MIN X MIN 1/4) - (Y MAX X MAX 1/4)
         IF CRIT%(NO)=10 THEN Y=Y+X+1/2-(Y MIN X MIN 1/2) - (Y MAX X MAX 1/2)
         IF CRIT%(NO)=11 THEN Y=(Y+X)/2
         IF CRIT%(NO)=13THEN Y=Y+X+3/4-(Y MIN X MIN 3/4) - (Y MAX X MAX 3/4)
```

```
        IF CRIT%(NO)=14 THEN Y=Y MAX X
        IF CRIT%(NO)=15 THEN Y=Y+X-Y*X
        IF CRIT%(NO)=16  THEN Y=1 MIN (Y+X)
      NEXT J
      SOMMET'PILE=SOMMET'PILE-1
      RETURN


REM List of response triples stored in the memory
DATA 1,1,3,1,2,3,1,3,3,1,3,4,2,3,3,3,3,3,3,3,4,3,3,5,4,3,4,5,3,5,5,4,5,5,5,5
DATA "CAR A", 21000,20,110,6.2
DATA "CAR B", 24500,18,130,6

      2.
REM ==========================================================================
REM =                                                                        =
REM =                 SUBROUTINE : COMPARISON OF FUZZY NUMBERS               =
REM =                                                                        =
REM = This subroutine computes the four indices introduced for the comparison=
REM = of fuzzy numbers X and Y, Y being used as the reference.              =
REM =                                                                        =
REM = Parameters : X, Y, PSE, NSE, PS, NS                                    =
REM = Output : PSE possibility that X is greater than or equal to Y         =
REM =          NSE necessity     "  "  "   "        "   "   "    " "       =
REM =          PS  possibility    "  "  "   "        "   "  Y             =
REM =          NS  necessity      "  "  "   "        "   "               =
REM ==========================================================================
COMP'NBF:
MI=VMI(X) : MS=VMS(X) : ALP=ALPHA(X) : BET=BETA(X)
NI=VMI(Y) : NS=VMS(Y) : GAM=ALPHA(Y) : DEL=BETA(Y)

REM When a spread is zero, a very small fictitious spread EPS is introduced
REM in order to be able to apply the formulas directly.
REM


EPS =.000001
EPS1=ABS(MI-NI) : IF EPS1=0 THEN EPS1=EPS
EPS2=ABS(MS-NS) : IF EPS2=0 THEN EPS2=EPS
EPS3=ABS(MI-NS) : IF EPS3=0 THEN EPS3=EPS
EPS4=ABS(MS-NI) : IF EPS4=0 THEN EPS4=EPS
EPS=(EPS MIN EPS1 MIN EPS2 MIN EPS3 MIN EPS4)/4
IF ALP=0 THEN ALP=EPS
IF BET=0 THEN BET=EPS
IF GAM=0 THEN GAM=EPS
IF DEL=0 THEN DEL=EPS

REM Computation of the four indices :
REM ---------------------------------

PSE=1+(MS-NI)/(GAM+BET)
PSE=0 MAX (1 MIN PSE)
PS=(MS-NS+BET)/(BET+DEL)
PS=0 MAX (1 MIN PS)
IF MS=NS AND BET=DEL THEN PS=0
NSE=(MI-NI+GAM)/(ALP+GAM)
NSE=0 MAX (1 MIN NSE)
IF MI=NI AND ALP=GAM THEN NSE=1
NS=(MI-NS)/(ALP+DEL)
NS=0 MAX (1 MIN NS)
RETURN
```

References

1. Aczel, J. (1966). *Lectures on Functional Equations and Their Applications.* Academic, New York.
2. Baas, S., and Kwakernaak, H. (1977). Rating and ranking of multiple aspect alternatives using fuzzy sets. *Automatica,* **13,** 47–58.

3. BALDWIN, J. F., and GUILD, N. C. F. (1979). Comparison of fuzzy sets on the same decision space. *Fuzzy Sets and Syst.*, **2**(3), 213–233.
4. BELLMAN, R. E., and ZADEH, L. A. (1970). Decision-making in a fuzzy environment. *Manage. Sci.*, **17**, B141–B164.
5. DOMBI, J. (1982). Basic concepts for a theory of evaluation: The aggregative operator. *Eur. J. Oper. Res.*, **10**, 282–293.
6. DUBOIS, D. (1983). Modèles mathématiques de l'imprécis et de l'incertain en vue d'applications aux techniques d'aide à la décision. Thesis, University of Grenoble.
7. DUBOIS, D., and PRADE, H. (1980). New results about properties and semantics of fuzzy set-theoretic operators. In *Fuzzy Sets: Theory and Applications to Policy Analysis and Information Systems* (P. P. Wang, and S. K. Chang, eds). Plenum Press, New York, pp. 59–75.
8. DUBOIS, D., and PRADE, H. (1983). The use of fuzzy numbers in decision analysis. In *Fuzzy Information and Decision Processes* (M. M. Gupta, and E. Sanchez, eds). North-Holland, Amsterdam, pp. 309–321.
9. DUBOIS, D., and PRADE, H. (1983). Ranking fuzzy numbers in the setting of possibility theory. *Inf. Sci.*, **30**(2), 183–224.
10. DUBOIS, D., and PRADE, H. (1983). Two-fold fuzzy sets: An approach to the representation of sets with fuzzy boundaries, based on possibility and necessity measures. *Fuzzy Math.* (*China*), **3**(4), 53–76.
11. DUBOIS, D., and PRADE, H. (1984). Criteria aggregation and ranking of alternatives in the framework of fuzzy set theory. *Fuzzy Sets and Decision Analysis* (H.-J. Zimmermann, L. A. Zadeh, and B. R. Gaines, eds). TIMS Studies in Management Sciences, Vol. 20, North-Holland, Amsterdam, pp. 209–240.
12. FRANK, M. J. (1979). On the simultaneous associativity of $F(x, y)$ and $x + y - F(x, y)$. *Aequationes Math.*, **19**, 194–226.
13. FREELING, A. N. S. (1980). Fuzzy sets and decision analysis. *IEEE Trans. Syst. Man Cybern.*, **10**, 341–354.
14. FUNG, L. W., and FU, K. S. (1975). An axiomatic approach to rational decision-making in a fuzzy environment. In *Fuzzy Sets and Their Applications to Cognitive and Decision Processes* (L. A. Zadeh *et al.*, eds). Academic, New York, pp. 227–256.
15. HAMACHER, H. (1975). Über logische Verknupfungen Unscharfer Aussagen und deren Zugehörige Bewertungs-funktionen. In *Progress in Cybernetics and Systems Research*, vol. 3 (R. Trappl, G. J. Klir, and L. Ricciardi, eds). Hemisphere, New York, pp. 276–287.
16. KANDEL, A., and BYATT, W. J. (1978). Fuzzy sets, fuzzy algebra and fuzzy statistics. *Proc. IEEE*, **68**, 1619–1639.
17. KEENEY, R. L., and RAIFFA, H. (1976). *Decisions With Multiple Objectives: Preferences and Value Trade-offs*. Wiley, New York.
18. LING, C. H. (1965). Representation of associative functions. *Publ. Math. Debrecen*, **12**, 189–212.
19. MENGER, K. (1942). Statistical metrics, *Proc. Natl. Acad. Sci. USA*, **28**, 535–537.
20. OKUDA, T., TANAKA, H., and ASAI, K. (1978). A formulation of fuzzy decision problems with fuzzy information, using probability measures of fuzzy events, *Inf. Control*, **38**(2), 135–147.
21. ROY, B. (1978). ELECTRE III: Un algorithme de classement fondé sur une représentation floue des préférences en présence de critéres multiples. *Cah. CERO*, **20**, 3–24.
22. SAATY, T. L. (1978). Exploring the interfaces between hierarchies, multiple objectives, and fuzzy sets, *Fuzzy Sets Syst.* **1**, 57–68.

23. SCHWEIZER, B., and SKLAR, A. (1963). Associative functions and abstract semigroups. *Publ. Math. Debrecen,* **10,** 69–81.
24. SILVERT, W. (1979). Symmetric summation: A class of operations on fuzzy sets. *IEEE Trans. Syst., Man. Cybern.,* **9,** 657–659.
25. SISKOS, J., LOCHARD, J., and LOMBARD, J. (1984). A multicriteria decision-making methodology under fuzziness: Application to evaluation of radiological protection in nuclear power-plants. In *Fuzzy Sets and Decision Analysis* (H. J. Zimmermann, L. A. Zadeh, and B. B. Gaines, eds). TIMS Studies in the Management Sciences, Vol. 20, North-Holland, Amsterdam, pp. 261–283.
26. THÖLE, U., ZIMMERMANN, H. J., and ZYSNO, P. (1979). On the suitability of minimum and product operators for the intersection of fuzzy sets. *Fuzzy Sets Syst.,* **2,** 167–180.
27. TRILLAS, E. (1979). Sobre funciones de negacion en la teoria de conjuntos difusos. *Stochastica,* **III**(1), 47–59.
28. TSUKAMOTO, Y., NIKIFORUK, P. N., and GUPTA, M. M. (1981). On the comparison of fuzzy sets using fuzzy chopping. *Proc. 8th Triennal World Congress IFAC,* Kyoto, Vol. 5, pp. 46–52.
29. VON NEUMANN, J., and MORGENSTERN, O. (1944). *Theory of Games and Economic Behavior.* Princeton Univ. Press, Princeton, New Jersey.
30. WATSON, S. R., WEISS, J. J., and DONNELL, M. (1979). Fuzzy decision analysis. *IEEE Trans. Syst. Man. Cybern.,* **9,** 1–9.
31. YAGER, R. R. (1977). Multiple-objective decision-making using fuzzy sets. *Int. J. Man-Machine Stud.,* **9,** 375–382.
32. YAGER, R. R. (1978). Fuzzy decision-making including unequal objectives. *Fuzzy Sets Syst.,* **1,** 85–95.
33. YAGER, R. R. (1980). On a general class of fuzzy connectives. *Fuzzy Sets Syst.,* **4,** 235–242.
34. YAGER, R. R. (1982). Some procedures for selecting fuzzy set-theoretic operators. *Int. J. Gen. Syst.,* **8**(2), 115–124.
35. ZIMMERMANN, H. J., and ZYSNO, P. (1980). Latent connectives in human decision making. *Fuzzy Sets Syst.,* **4**(1), 37–51.
36. ZIMMERMANN, H. J., and ZYSNO, P. (1983). Decisions and evaluations by hierarchical aggregation of information. *Fuzzy Sets Syst.,* **10,** 243–260.
37. ZYSNO, P. (1982). The integration of concepts within judgmental and evaluative processes. *Progress in Cybernetics and Systems Research,* vol. VIII (R. Trappl, G. Klir, and F. Pichler, eds), Hemisphere, New York, pp. 509–517.
38. DUBOIS, D., and PRADE, H. (1985). A review of fuzzy set aggregation connectives. *Inf. Sci.,* **36,** 85–121.
39. DUBOIS, D., and PRADE, H. (1986). Weighted minimum and maximum operations in fuzzy set theory. An addendum to "A review of fuzzy set aggregation connectives." *Inf. Sci.* **39,** 205–210.
40. YAGER, R. R. (1981). A new methodology for ordinal multiple aspect decisions based on fuzzy sets. *Decision Sci.,* **12,** 589–600.
41. YAGER, R. R. (1984). General multiple objective decision functions and linguistically quantified statements. *Int. J. Man-Machine Stud.,* **21,** 389–400.
42. BORTOLAN, G., and DEGANI, R. (1985). A review of some methods for ranking fuzzy numbers. *Fuzzy Sets Syst.* **15,** 1–19.
43. ROUBENS, M., VINCKE, P. (1988). Fuzzy possibility graphs and their application to ranking fuzzy numbers. *Fuzzy Sets and Systems,* to be published.

4

Models for Approximate Reasoning in Expert Systems

In the expert systems of artificial intelligence, the facts and/or the rules to be represented may often be uncertain or imprecise.

For a long time, probability theory was the only available numerical approach to the problem of uncertain inference. Many mathematical models for uncertainty, differing distinctly from probability, have recently been proposed, especially Shafer's theory of belief functions and possibility theory. At the same time, investigators in the field of artificial intelligence have felt the need for alternatives to the standard Bayesian model and have proposed more empirical models such as those in, especially, MYCIN [40] and PROSPECTOR [15] (see also [17, 21, 23]).

In this chapter we present a synthetic view of several theoretically based, and not purely probabilistic, deductive approaches. The first section rounds off the material of Chapter 1 concerning various models of imprecision and uncertainty which generalize the concepts of probability and possibility. Then, in the succeeding two sections, follows a treatment of the two principal inference mechanisms needed in expert systems, namely, deductive inference and the combination of data derived from different sources, in the context of uncertain or imprecise premises.

4.1. Remarks on Modeling Imprecision and Uncertainty

In this chapter, an item of information is represented as a logical proposition denoted p, q, or r. Set-theoretic notation is only used when we wish to display the content (possibly imprecise) of these propositions. Thus we shall consider the set $\mathscr{P}$ of propositions such that:

1. If $p \in \mathscr{P}$, then $\neg p \in \mathscr{P}$ ($\neg$ = negation).
2. If $p \in \mathscr{P}$ and $q \in \mathscr{P}$, then $p \wedge q \in \mathscr{P}$ ($\wedge$ = conjunction).

We denote by $\mathbb{0}$ the necessarily false proposition, and by $\mathbb{1}$ the necessarily true position. Clearly, $\mathbb{0} \in \mathscr{P}$, $\mathbb{1} \in \mathscr{P}$. In fact, $\mathscr{P}$, furnished with the operators $\neg$, $\wedge$, and $\vee$ [disjunction, defined by $p \vee q = \neg(\neg p \wedge \neg q)$] is a Boolean lattice of *classical* logical propositions. Implication is denoted by $\rightarrow$ and is defined as usual by $p \rightarrow q = \neg p \vee q$. We say that "$p$ entails q" when $p \rightarrow q = \mathbb{1}$. When $p \wedge q = \mathbb{0}$, we say that "p and q are incompatible" since if one of the propositions is true, the other is false; also, "$p \wedge q = \mathbb{0}$" is equivalent to "$(p \rightarrow \neg q) = \mathbb{1}$," i.e., to "$p$ entails not-q."

4.1.1. Credibility and Plausibility

Here we suppose that $\mathscr{P}$ is a finite set.

In probability theory, the probability $P(p)$ that p is true and the probability $P(\neg p)$ that p is false are related by $P(p) + P(\neg p) = 1$. Thus if $P(p) = 0$ then $P(\neg p) = 1$; if nothing is known *a priori* as to the truth or falsity of p, it seems natural to take $P(p) = P(\neg p) = \frac{1}{2}$. However, once one has more than two alternatives (incompatible by pairs), "complete ignorance" is difficult to represent, for whatever the probability distribution certain propositions (other than $\mathbb{1}$) will be more probable than certain others (other than $\mathbb{0}$), which is paradoxical in supposedly complete ignorance (see Chapter 1, Section 2).

In the 1970s, several lines of research led to "measures of uncertainty" being proposed that were different from probability, but that had in common the following minimal and intuitively reasonable properties. Let g be such a measure of uncertainty on $\mathscr{P}$, taking values in $[0, 1]$:

1. $g(\mathbb{0}) = 0$
2. $g(\mathbb{1}) = 1$ \hfill (4.1)
3. If p implies q, then $g(q) \geqslant g(p)$

These are the "confidence measures" considered in Chapter 1, Section 1.3.

However, the axioms (4.1) characterize a family of confidence measures that is too large, and unsuited to computation.

Shafer [39] introduced measures of credibility or belief [which naturally satisfy (4.1)] which can be expressed in terms of a function m on $\mathscr{P}$ taking values in $[0, 1]$, such that

$$m(\mathbb{0}) = 0 \qquad \sum_{p \in \mathscr{P}} m(p) = 1 \qquad (4.2)$$

The measure of credibility Cr, based on m, is then expressed by

$$\forall q \in \mathscr{P}, \qquad \mathrm{Cr}(q) = \sum_{p \text{ entails } q} m(p) \qquad (4.3)$$

$m(p)$ represents the "quota of belief" associated with p and with p alone, $\mathrm{Cr}(q)$—*the credibility of* q—being obtained as the sum of the quotas of belief in the propositions that entail q. The measure m is *not* a confidence measure [it fails to satisfy (4.1)], but a relative weighting. Propositions p such that $m(p) > 0$ are called "focal propositions."

Dually, one can then define, in terms of Cr, a measure Pl of plausibility:

$$\forall p \in \mathscr{P}, \qquad \mathrm{Pl}(p) = 1 - \mathrm{Cr}(\neg p) \qquad (4.4)$$

which, in terms of m, gives

$$\forall q \in \mathscr{P}, \qquad \mathrm{Pl}(q) = \sum_{p \text{ does not entail} \neg q} m(p) \qquad (4.5)$$

Shafer [39] shows that the functions Cr and Pl, respectively, satisfy (4.3) and (4.5) for a weighting m in the sense of (4.2), if and only if they are, respectively, superadditive and subadditive to every positive integer order n. To the second order, these properties are

$$\mathrm{Cr}(p \vee q) \geqslant \mathrm{Cr}(p) + \mathrm{Cr}(q) - \mathrm{Cr}(p \wedge q) \qquad \text{(superadditivity)}$$
$$\mathrm{Pl}(p \wedge q) \leqslant \mathrm{Pl}(p) + \mathrm{Pl}(q) - \mathrm{Pl}(p \vee q) \qquad \text{(subadditivity)}$$

It then follows that

$$\forall p \in \mathscr{P}, \qquad \mathrm{Cr}(p) + \mathrm{Cr}(\neg p) \leqslant 1 \qquad (4.6)$$
$$\forall p \in \mathscr{P}, \qquad \mathrm{Pl}(p) + \mathrm{Pl}(\neg p) \geqslant 1 \qquad (4.7)$$

Thus it is possible to have $\mathrm{Cr}(p) = \mathrm{Cr}(\neg p) = 0$ and $\mathrm{Pl}(p) = \mathrm{Pl}(\neg p) = 1$ in the case of complete ignorance: two contradictory propositions can appear plausible without either of them being in the least credible. Further, (4.3) and (4.5) show clearly that

$$\forall p \in \mathcal{P}, \qquad \mathrm{Cr}(p) \leqslant \mathrm{Pl}(p) \tag{4.8}$$

The plausibility of a proposition is always at least as great as its credibility, which seems intuitively satisfying just as does the relation (4.4), which means that a proposition is the more credible according as its contrary is less plausible.

Note that, *if every focal proposition is incompatible with every proposition not entailed by it, then the measures of credibility and plausibility, defined by (4.3) and (4.5), are the same.* This condition can be written formally

$$\forall p \in \{p \mid m(p) > 0\}, \forall q, \text{ if } p \to q \neq \mathbb{1}, \text{ then } p \wedge q = \mathbb{0} \tag{4.9}$$

The identity of Pl and Cr, along with subadditivity of Pl and superadditivity of Cr, show that in this case the confidence measure is a probability measure P.

Let an "elementary proposition" be one that is entailed only by itself (or any equivalent proposition) and by the universally false proposition, i.e., p such that:

$$\forall q \neq \mathbb{0}, \qquad \text{if } q \leftrightarrow p \neq \mathbb{1} \qquad \text{then } q \to p \neq \mathbb{1} \tag{4.10}$$

where $\leftrightarrow$ stands for equivalence [i.e., $p \leftrightarrow q = (p \to q) \wedge (q \to p)$]. It can be shown that (4.9) implies that every focal proposition is elementary, and conversely. The property $\mathrm{Pl} = \mathrm{Cr} = P$ is therefore equivalent to the fact that focal propositions are all elementary (and therefore incompatible). This result is the expression, in logical terms, of the discussion given in Chapter 1, Section 1.3.2 on the specificity of probability measures.

Apart from probability, there is another important class of credibility and plausibility measures: the measures of *necessity* and of *possibility*.

Suppose the n focal propositions are *coherent*, that is to say that they can be ordered so that p_n entails p_{n-1} entails $\cdots$ entails p_1. Then it can be shown that

$$\forall p \in \mathcal{P}, \qquad \forall q \in \mathcal{P}, \qquad \mathrm{Cr}(p \wedge q) = \min(\mathrm{Cr}(p), \mathrm{Cr}(q)) \tag{4.11}$$

$$\forall p \in \mathcal{P}, \qquad \forall q \in \mathcal{P}, \qquad \mathrm{Pl}(p \vee q) = \max(\mathrm{Pl}(p), \mathrm{Pl}(q)) \tag{4.12}$$

The measures of necessity and of possibility can be recognized here as special cases of the measures of credibility and of plausibility, respectively (see Chapter 1); their axioms are here expressed in terms of propositions rather than events. Recall the property

$$\forall p \in \mathcal{P}, \begin{cases} \Pi(p) < 1 \Rightarrow N(p) = 0 \\ N(p) > 0 \Rightarrow \Pi(p) = 1 \end{cases} \qquad (4.13)$$

which means that a *classical* proposition (i.e., $p \wedge \neg p = \mathbb{0}$ and $p \vee \neg p = \mathbb{1}$) is always completely possible before in the least necessary (= certain).

N.B.1. An event with probability equal to 1 is regarded as certain; this is not so for an event with possibility equal to 1 since the contrary event can also have possibility equal to 1. On the other hand, if the necessity of an event is 1, it can be taken as certain, since the necessity and the possibility of the contrary even are both zero. $\qquad \square$

N.B.2. Given a weighting m, in the sense of (4.2), on the focal propositions of $\mathcal{P}$, let σ be a function under which to each focal proposition p corresponds one elementary proposition $\sigma(p)$ such that $\sigma(p) \to p = \mathbb{1}$, and let P_σ be a probability measure generated by a weighting m_σ such that

$$\forall q, \quad m_\sigma(q) = m(p) \quad \text{if } q = \sigma(p)$$
$$= 0 \quad \text{otherwise}$$

then it can be verified that $\forall \sigma, \forall p \in \mathcal{P}, \ \mathrm{Cr}(p) \leqslant P_\sigma(p) \leqslant \mathrm{Pl}(p)$ [cf. Chapter 1, Section 1.6; here we find again (1.56)]. This is in accordance with the intuitive requirement that what is credible must be probable, and what is probable must be plausible. $\qquad \square$

4.1.2. Decomposable Measures

Another context in which measures of probability, possibility, and necessity are found is that of decomposable confidence measures g (Dubois and Prade [8, 10]), which satisfy the following axioms (where $\mathcal{P}$ is supposed to be finite):

$$g(\mathbb{0}) = 0, \quad g(\mathbb{1}) = 1$$
$$\exists \perp, \quad p \wedge q = \mathbb{0} \Rightarrow g(p \vee q) = g(p) \perp g(q) \qquad (4.14)$$

where $[0, 1]$ is closed under $\perp$: (4.14) is a natural hypothesis which says that the degree of confidence in the proposition "p or q" depends only on the separate degrees of confidence in p and in q when p and q are incompatible. The Boolean lattice structure of $\mathcal{P}$, with the monotonic character of g [equation (4.1)], leads to $\perp$ being chosen only from the triangular co-norms introduced in Section 3.1.2.2 of Chapter 3. In particular, if we choose $\perp = \max$, we find that g is a possibility measure; if $\perp$ is the bounded sum $[u * v = \min(1, u + v)]$ and we impose the normalization condition

$$\sum \{g(p) \mid p \text{ is elementary}\} = 1$$

then g is a probability measure. Decomposable measures have moreover the property that they can be completely defined in terms of a co-norm $\perp$ and their values over elementary propositions, since any proposition can be written as a disjunction of the elementary propositions that entail it.

Any decomposable measure g has a dual measure g_c defined by

$$\forall p, \qquad g_c(p) = c(g(\neg p))$$

where c is a complementation operator (Chapter 3, Section 3.1.2.1). Then g_c is constructed from the triangular norm $*$ such that $u * v = c(c(u) \perp c(v))$, and satisfies an axiom that is dual to (4.14):

$$\forall p, q, \qquad p \vee q = \mathbb{1} \Rightarrow g_c(p \wedge q) = g_c(p) * g_c(q) \qquad (4.15)$$

which is reminiscent of the axiom for necessity measures (obtained with (4.15) and $* = \min$). Noting that (4.14) implies $g(p) \perp g(\neg p) = 1$, decomposable measures can be grouped into two classes:

- Those where knowledge of $g(p)$ completely defines the value of $g(\neg p)$. These confidence measures are called self-dual ($\exists c$, $g_c = g$) and satisfy both (4.14) and (4.15) together. Their prototype is the probability measure.
- Those where $g(\neg p)$ cannot always be deduced from $g(p)$. This is particularly the case with confidence measures arising from the max operator or from a strict co-norm (Chapter 3, Section 3.1.2.2) such as $u \perp v = u + v - uv$; they satisfy $\max(g(p), g(\neg p)) = 1$, which makes them similar to the possibility measures. They are in general different from their duals, which are similar to the necessity measures.

For a more detailed exposition of this class of measures, see Dubois and Prade [8] and Weber [68].

4.1.3. Vague Propositions

The content of a proposition p of the form "X takes its value over A" or, more briefly, "X is A," specifies precisely or imprecisely, the value of a variable X in a universe of discourse S by means of a predicate A which corresponds to a subset of S. In particular, every elementary proposition $p \in \mathcal{P}$ is of the form "X takes the value s" where $s \in S$; we can denote it by p_s. Such a proposition is called *precise* with respect to the reference set S. Every nonelementary proposition (if other than $\mathbb{0}$) is therefore imprecise with respect to this reference set. In fact, we are here identifying $(\mathcal{P}, \neg, \wedge, \vee)$ with $(\mathcal{P}(S), -, \cap, \cup)$, where $\mathcal{P}(S)$ is the set of parts of S. From this point of view, $\mathbb{0} = $ "X is $\varnothing$" means that "X does not take any value over S" (which is of course necessarily false since by definition X takes its value in S), while $\mathbb{1}$ corresponds to the trivial affirmation that "X takes its value over S." Note that in this chapter the elements of S are supposed to be mutually exclusive values which can be assigned to X.

Given a possibility measure Π which evaluates the uncertainty of the propositions in $\mathcal{P}$, we can write [cf. formula (1.9)]

$$\forall p, \; \Pi(p) = \sup\{\Pi(p_s) \mid p_s \to p = \mathbb{1}\} \tag{4.16}$$

The possibility distribution $\{\Pi(p_s) \mid s \in S\}$, which characterizes Π, can be interpreted as a vague proposition of the form "X is A," where A is a (normalized) fuzzy subset of S defined by

$$\mu_A(s) = \Pi(p_s) \triangleq \pi_X(s)$$

The notation π_X exhibits the variable associated with the proposition. In particular, if $\forall p, \; \Pi(p) \in \{0, 1\}$, Π is equivalent to the classical proposition $q = \vee \{p_s \mid s \in A\}$ with $A = \{s \mid \pi_X(s) = 1\}$. Specifying a possibility distribution π_X over S corresponds as well, in practice, to the representation of a vague predicate ("tall," "young," "heavy") as to the setting out of a list of mutually exclusive alternatives weighted with their degrees of possibility (without any explicit preexisting predicate); moreover, as we have seen, π_X is equally equivalent to giving a set of coherent imprecise alternatives, weighted by values that can be regarded as degrees of probability [cf. Chapter 1, Section 1.4, Equation (1.42)].

4.1.4. Evaluating the Truth Value of a Proposition

The truth value of a proposition can be regarded as a measure of the extent to which its content agrees with the content of our knowledge of reality (which may be incomplete). The content of the proposition "X is F" to be evaluated, and the content of the reference proposition "X is A", are represented by the possibility distributions μ_F and μ_A, respectively; these express the constraints placed by the propositions on the values of the variable X. The possibility, and the necessity, that the proposition "X is F" should be true, given the knowledge that "X is A," can be evaluated as

$$\Pi(F;A) = \sup_{s \in S} \min(\mu_F(s), \mu_A(s)) = \Pi(A;F) \qquad (4.17)$$

$$N(F;A) = \inf_{s \in S} \max(\mu_F(s), 1 - \mu_A(s)) \qquad (4.18)$$

which are, respectively, the possibility and the necessity of the fuzzy event F calculated in terms of the possibility distribution $\pi_X = \mu_A$ (Chapter 1, Section 1.7). Observe that when our knowledge is precise (i.e., complete) and therefore A corresponds to a singleton of S, then (4.17) and (4.18) give

$$\Pi(F;A) = N(F;A) \qquad (4.19)$$

while if F is not fuzzy (i.e., "X is F" is not a vague proposition), then whatever the nature of A

$$\Pi(F;A) = 1 \quad \text{or else} \quad N(F;A) = 0 \qquad (4.20)$$

We recover the truth values of classical logic when our knowledge corresponds to a singleton $\{s_0\}$ of S, and the evaluated propositions are not vague. Then, if $p = $ "X is F,"

$$v(p) = \Pi(F;\{s_0\}) = N(F;\{s_0\}) = \mu_F(s_0) \in \{0, 1\}$$

With a Cartesian product $S \times T$ of reference sets, if the variables X and Y are noninteractive, the reference knowledge can be represented by the Cartesian product $A \times B$ of fuzzy sets [cf. (1.72)], and the degrees of possibility and of necessity of the vague propositions corresponding to the events $F \times G$ and $F + G$ [where "$+$" is a Cartesian coproduct; cf. (1.73)] satisfy the following equalities already given in Chapter 1, Section 1.8:

$$\Pi(F \times G; A \times B) = \min(\Pi(F;A), \Pi(G, B)) \qquad (4.21)$$

$$N(F \times G; A \times B) = \min(N(F;A), N(G;B)) \qquad (4.22)$$

$$\Pi(F + G; A \times B) = \max(\Pi(F;A), \Pi(G;B)) \qquad (4.23)$$

$$N(F + G; A \times B) = \min(N(F;A), N(G;B)) \qquad (4.24)$$

N.B. So that $\Pi(F; A)$ or $N(F; A)$ may be considered as a degree of truth $v(F; A)$ in the sense of an extensional logic, we would wish to have the following equalities, which hold in classical logic:

$$v(F; A) + v(\bar{F}; A) = 1 \qquad (4.25)$$

$$v(F \cap G; A) = f(v(F; A), v(G; A)) \qquad (4.26)$$

$$v(F \cup G; A) = g(v(F; A), v(G; A)) \qquad (4.27)$$

Clearly neither $\Pi(F; A)$ nor $N(F; A)$ satisfies (4.25) in general. Nonetheless, if A is a singleton then extensionality is preserved for these two quantities on taking $f = \min$ and $g = \max$, and they are then equal [cf. (4.19)]. Further, (4.25) is still true in general if, following Gaines [18], we put

$$v(F; A) = \frac{\Pi(F; A) + N(F; A)}{2} \qquad (4.28)$$

But, if F and G are standard sets, then v, defined as above, satisfies (4.26) and (4.27) when F and G are on distinct reference sets S and T, A is replaced by a Cartesian product $A \times B$ on $S \times T$, and $F \cup G$ (or, respectively, $F \cap G$) is replaced by $F + G$ (or, respectively, $F \times G$); this result holds by virtue of the equalities (4.21)–(4.24). See Dubois and Prade [9] for a more detailed discussion. □

The degree of conformity of "X is F" with "X is A" is more completely evaluated by the compatibility function $\text{CP}(F; A)$ (see Zadeh [51, 52]), which is a fuzzy subset of the interval $[0, 1]$ defined, using the extension principle (2.6), by

$$\mu_{\text{CP}(F;A)}(v) = \sup_{s, v = \mu_F(s)} \mu_A(s)$$

$$= 0 \qquad \text{if } \mu_F^{-1}(v) = \varnothing \qquad (4.29)$$

The fuzzy subset $\text{CP}(F; A)$ of $[0, 1]$ already encountered in Chapter 3 Section 3.2 is none other than the fuzzy subset of possible values of the degree of F-membership of an element whose set of a priori possible values on S is restricted by A. In other words, given that s is a more or less representative element of A, $\text{CP}(F; A)$ is the possibility distribution of the variable $\mu_F(s)$. Note that if F is nonfuzzy, $\text{CP}(F; A)$ is a fuzzy subset of $\{0, 1\}$ such that

$$\mu_{\text{CP}(F;A)}(0) = \Pi(\bar{F}; A), \qquad \mu_{\text{CP}(F;A)}(1) = \Pi(F; A) \qquad (4.30)$$

If A is a singleton $\{s\}$, $CP(F; A) = \mu_F(s)$. If A is a classical subset then $CP(F; A)$ is a subset of $[0, 1]$ bounded above by $\Pi(F; A)$ and below by $N(F; A)$.

In this general case, it can be shown (see, e.g., [32]) that

$$\Pi(F; A) = \sup_{v \in [0,1]} \min(v, \mu_{CP(F;A)}(v)) \tag{4.31}$$

$$N(F; A) = \inf_{v \in [0,1]} \max(v, 1 - \mu_{CP(F;A)}(v)) \tag{4.32}$$

Thus $\Pi(F; A)$ and $N(F; A)$ are implicit in $CP(F; A)$. By means of operations on fuzzy intervals (Chapter 2), an expression for $CP(\bar{F}; A)$ can easily be obtained from $CP(F; A)$ and $CP(F \times G; A \times B)$ or $CP(F + G; A \times B)$ in terms of $CP(F; A)$ and $CP(G; B)$ when there is noninteraction between the variable X associated with F and A and the variable Y associated with G and B; we then arrive at the analog of the equalities (4.21)–(4.24).

The "truth" of a proposition has been calculated as the compatibility of the possibility distribution representing the proposition with the possibility distribution representing the pertinent state of knowledge. Now we turn to the converse problem: given a proposition p qualified by a truth value τ, we seek that state of knowledge with which p is compatible to degree τ. In other words, given a proposition of the form "p is τ," where p is itself a proposition and τ represents a (possibly fuzzy) truth value, we seek to represent the proposition "p is τ" in the form of a possibility distribution which is compatible to degree τ with that associated with p.

Given a possibility distribution μ_F representing $p = $ "X is F," and the characteristic function μ_τ of the fuzzy subset of $[0, 1]$ representing the truth value τ, the maximal solution μ_A^+ (in the sense of fuzzy-set inclusion) of the equation $\tau = CP(F; A)$, where $CP(F; A)$ is defined by (4.29), is given (see Zadeh [52]) by

$$\forall s \in S \qquad \mu_A^+(s) = \mu_\tau(\mu_F(s)) \tag{4.33}$$

μ_A^+ represents what may be concluded about reality when it is known that "X is F" is "τ-true."

Remark: Measures of belief and disbelief used in MYCIN.

A measure of belief $MB(h, e)$ and a measure $MD(h, e)$ of disbelief in the hypothesis h given evidence e were introduced empirically in the expert system MYCIN [40]. They satisfy the following properties:

$$MB(\neg h, e) = MD(h, e) \tag{4.34}$$

and

$$1 - MD(h, e) < 1 \Rightarrow MB(h, e) = 0$$
$$MB(h, e) > 0 \Rightarrow MD(h, e) = 0 \tag{4.35}$$

where h is the proposition to be evaluated and e is the reference proposition. Equations (4.34) and (4.35) are the analog of (4.4) and (4.13), respectively, where $MB(h, e)$ behaves like a necessity measure and $MD(h, e)$ like the complement with respect to 1 of a possibility measure, as is remarked in [33]. Moreover, the following formulas used in MYCIN

$$MD(h_1 \wedge h_2, e) = \max(MD(h_1, e), MD(h_2, e)) \tag{4.36}$$

$$MB(h_1 \wedge h_2, e) = \min(MB(h_1, e), MB(h_2, e)) \tag{4.37}$$

$$MD(h_1 \vee h_2, e) = \min(MD(h_1, e), MD(h_2, e)) \tag{4.38}$$

$$MB(h_1 \vee h_2, e) = \max(MB(h_1, e), MB(h_2, e)) \tag{4.39}$$

are the exact counterparts of formulas (4.21)–(4.24).

The certainty factor

$$CF(h, e) = MB(h, e) - MD(h, e) \in [-1, +1]$$

used in MYCIN can be considered as a truth value to within linear transformation, since $[1 + CF(h, e)]/2$ is the analog of the quantity defined by (4.28), and we also have $CF(h, e) + CF(\neg h, e) = 0$, which is the analog of (4.25). $\qquad\qquad\qquad\qquad\qquad\qquad\qquad\qquad\qquad \square$

4.2. Reasoning from Uncertain Premises

In the context of automated reasoning, we can apparently adopt two approaches to knowledge representation: the logical approach and the "functional approach."

In the logical approach, items of knowledge—facts and rules—are represented as logical assertions, and inference is based on the use of

TABLE 4.1. Modus Ponens

$v(p \rightarrow q)\, v(p)$		$v(q)$	
1	1	1	$\leftarrow$ Modus ponens
1	0	$\{0, 1\}$	$\leftarrow$ "Denial of q"; however, its truth value is indeterminate.
0	1	0	
0	0	$\varnothing$	$\leftarrow$ Impossible situation

detachment rules. In classical logic, the two most heavily used rules are the following

- *Modus ponens*:

$$p \rightarrow q$$
$$\underline{p\qquad}$$
$$q$$

corresponding to the first line of the inverted truth table in Table 4.1, where $v(p)$ is the truth value of the proposition p.
- *Modus tollens*:

$$p \rightarrow q$$
$$\underline{\neg q}$$
$$\neg p$$

corresponding to the second line of the inverse truth table in Table 4.2.

TABLE 4.2. Modus Tollens

$v(p \rightarrow q)\, v(q)$		$v(p)$	
1	1	$\{0, 1\}$	$\leftarrow$ "Confirmation of p"; however, its truth value is indeterminate
1	0	0	$\leftarrow$ Modus tollens
0	1	$\varnothing$	$\leftarrow$ Impossible situation
0	0	1	

N.B. In Table 4.2, given that $p \to q$ is true, truth of q makes "p is true" more credible (see Polya [31]) in the sense that "q is true" is a necessary condition for p to be true; whence the impression of confirmation. Likewise, given that $p \to q$ is true, falseness of p makes "q is true" less plausible since it is a necessary condition for the falseness of q (or, if one likes, given that $p \to q$ is true, truth of p is sufficient for q to be true); whence the impression of denial. Nevertheless it is not possible in classical logic to quantify the gain in credibility nor the loss of plausibility.   □

The impossible cases in the above tables show that the truth values of p and of $p \to q$ cannot be chosen independently of each other. In the functional approach, the rules are viewed as partial specifications of functions whose arguments correspond to the facts. The reasoning is carried out by applying these functions to the available arguments. Thus the rules can be regarded as "precalculated inferences," which can be expressed in terms of conditionals when the validity of the rules is uncertain. The rule "if p then q" is interpreted as "q follows from p," where p and q are propositions of the form "$X \in A$" and "$Y \in B$," respectively. Let $g(q \mid p) \in \{0, 1\}$ be a binary relation on $\mathcal{P}$ defined by $g(q \mid p) = 1$ if and only if the rule "if p then q" is valid. In other words, $g(q \mid p) = 1$ means that q can be deduced from p. Caution: the symbol "$\mid$" is not a logical connective. In logic, when $g(q \mid p) = 1$, we normally write $p \vdash q$. Observe that if $g(q \mid p) = 0$ (i.e., q does not follow from p), or if $v(p) = 0$ (i.e., $X \notin A$), then no conclusion can be drawn as to the truth of q. A table similar to Table 4.1 can be drawn up, in which $g(q \mid p)$ has been substituted for $v(p \to q)$. In this case the same value for $v(q)$ occurs in line 1, and $\{0, 1\}$ (indeterminacy) in the other lines, which is equivalent to the inequalities

$$v(q) \geqslant v(p \wedge q) \geqslant \min(g(q \mid p), v(p)) \tag{4.40}$$

which means that $p \wedge q$ is true when p is true, if q can be inferred from p. Given the truth values $v(p)$ and $v(p \wedge q)$, a collection of relations g is implicitly defined by (4.40), and the least specific one satisfies the equality

$$v(p \wedge q) = \min(g(q \mid p), v(p)) \tag{4.41}$$

This approach has the advantage that the impossible case has been eliminated (line 4 of Table 4.1), which allows $v(p)$ and $g(q \mid p)$ to be defined independently, while $v(p)$ and $v(p \to q)$ are linked.

In the following, we shall discuss the problem of inference from uncertain premises using first the logical approach, and then the functional approach, obtaining similar results in both cases. Then we shall propose a procedure for inference from imprecise premises.

4.2.1. Deductive Inference with Uncertain Premises

In this section we suppose that the propositions in question are not vague, but the knowledge base, which could serve to establish their truth, is incomplete. As in Section 1.4, uncertainty about the truth of the proposition p is evaluated by means of a confidence measure which will be a possibility or a necessity measure. Uncertainty about the premise "if p then q" will be evaluated, depending on the point of view, as equal to that of the proposition $p \rightarrow q = \neg p \vee q$, or as the uncertainty about the conditioning of q by p.

4.2.1.1. *Modus Ponens and Modus Tollens with Uncertain Premises*

Let Π be a possibility measure on a Boolean lattice $\mathscr{P}$ of propositions, and let N be the dual necessity measure. In [34, 35, 14] we have proposed the following extensions of modus ponens and modus tollens:

Modus ponens   Modus tollens

(I)
$$\frac{N(p \rightarrow q) \geqslant a \quad N(p) \geqslant b}{N(q) \geqslant \min(a, b)}$$

(IV)
$$\frac{N(p \rightarrow q) \geqslant a \quad N(q) \leqslant b}{N(p) \leqslant 1 \text{ if } a \leqslant b \quad \leqslant b \text{ if } a > b}$$

(II)
$$\frac{N(p \rightarrow q) \geqslant a \quad \Pi(p) \geqslant b}{\Pi(q) \geqslant b \cdot v(a + b > 1)}$$

(V)
$$\frac{N(p \rightarrow q) \geqslant a \quad \Pi(q) \leqslant b}{\Pi(p) \leqslant \max(1 - a, b)}$$

(III)
$$\frac{\Pi(p \rightarrow q) \geqslant a \quad N(p) \geqslant b}{\Pi(q) \geqslant a \cdot v(a + b > 1)}$$
where $v(a + b > 1) = 1$ if $a + b > 1$
$= 0$ otherwise

- (I) and (IV) are obtained by noting that $p \wedge q = p \wedge (\neg p \vee q)$, which leads to $N(q) \geqslant N(p \wedge q) = \min(N(p \rightarrow q), N(p))$.
- (II) and (V) are obtained by noting that $\neg p \wedge \neg q = \neg q \wedge (\neg p \vee q)$, whence $N(\neg p) \geqslant N(\neg p \wedge \neg q) = \min(N(\neg q), N(p \rightarrow q))$, which finally gives $\Pi(p) \leqslant \max(\Pi(q), 1 - N(p \rightarrow q))$.
- (III) is easily verified since $\Pi(p \rightarrow q) = \max(1 - N(p), \Pi(q))$.

However, if we know only that $\Pi(p \to q) \geq a$, no conclusions can be drawn about $\Pi(p)$ from $\Pi(q) \leq b$, nor about $\Pi(q)$ from $\Pi(p) \geq b$; nor from $N(q) \leq b$.

Putting (I), (II), and (III) for modus ponens together, we get

$$
\text{(VI)} \quad
\begin{array}{llll}
N(p \to q) \geq a, & \Pi(p \to q) \geq A & (\max(1 - a, A) = 1) \\
N(p) \quad\;\; \geq b, & \Pi(p) \quad\;\; \geq B & (\max(1 - b, B) = 1) \\
\hline
N(q) \geq \min(a, b) & \Pi(q) \geq \max(A \cdot v(A + b > 1), & \\
 & \qquad\qquad\qquad B \cdot v(a + B > 1)) &
\end{array}
$$

where in every case $\max(\Pi(q), 1 - N(q)) = 1$.

In the case of probability measures, the rules of inference corresponding to (VI) (Suppes [43]), and to (IV) and (V), are well known:

$$
\text{(VII)} \quad
\begin{array}{ll}
P(p \to q) \geq a \\
P(q) \quad\;\; \geq b \\
\hline
P(q) \geq \max(0, a + b - 1)
\end{array}
\qquad
\text{(VIII)} \quad
\begin{array}{ll}
P(p \to q) \geq a \\
P(q) \quad\;\; \leq b \\
\hline
P(q) \leq \min(1, 1 - a + b)
\end{array}
$$

Schemes (IV) and (V) for modus tollens can also be combined in a similar way. A further schema can be made up by combining (I), (II), (IV), and (V) in the form

$$
\text{(IX)} \quad
\begin{array}{l}
N(p \to q) \geq a \\
N(q \to p) \geq a' \\
N(p) \in [b, b'], \; \Pi(p) \in [c, c'] \text{ and } \max(1 - b, c') = 1 \\
\hline
N(q) \in \left[\min(a, b), \begin{cases} 1 & \text{if } a' \leq b' \\ b' & \text{if } a' > b' \end{cases} \right] \\[2ex]
\Pi(q) \in [c \cdot v(a + c > 1), \max(1 - a', c')]
\end{array}
$$

In the last schema, reasoning proceeds according to a weighting on the equivalence between p and q. The magnitude a measures the extent to which p suffices to imply q, and a' measures to what extent p is necessary to imply q. Note that in all these schemas, if the coefficients a, b, a', b', c, c' are restricted to taking values 0 or 1, it turns out that the lower bounds for the evaluation of q are conjunctionlike, while the upper bounds are implicationlike. In (IX), the intervals containing $N(q)$ and $\Pi(q)$ include those containing $N(p)$ and $\Pi(p)$, respectively. These intervals do not increase in width if the procedure is reapplied, using in place of $N(p)$ and $\Pi(p)$ the intervals obtained for, respectively, $N(q)$ and $\Pi(q)$.

4.2.1.2. Conditioning on Uncertain Propositions [14]

The rule "if p then q" is a partial specification of a function from $\mathscr{P}$ into $\mathscr{P}$ in the form of a conditional possibility measure $\Pi(\cdot \mid p)$, where $\Pi(q \mid p)$ measures the possibility that q can be deduced from p. $\Pi(\cdot \mid p)$ is implicitly defined by the inequality, induced by (4.40),

$$\Pi(p \wedge q) \geq \Pi(q \mid p) * \Pi(p) \qquad (4.42)$$

once we have a possibility measure Π on $\mathscr{P}$ expressing our knowledge concerning the truth of the propositions. $*$ is an operator of conjunction type, such as the minimum, or more generally such that

1. if $r \leq s$ and $t \leq u$, then $r * t \leq s * u$ (monotonicity)
2. $\forall s \in [0, 1]$, $s * 0 = 0 * s = 0$
3. $\forall s \in [0, 1]$, $s * 1 = 1 * s = s$

For example, $*$ may be a continuous triangular norm such as min, the product, or else $s * t = \max(0, s + t - 1)$. The least specific possibility measure satisfying (4.42) attains the bounds of the inequality since $\Pi(p) \geq \Pi(p \wedge q)$. We can therefore define $\Pi(\cdot \mid p)$ by

$$\Pi(p \wedge r) = \Pi(r \mid p) * \Pi(p), \qquad r = q, \neg q \qquad (4.43)$$

Since p and q are classical propositions, $q = (p \wedge q) \vee (\neg p \wedge q)$, and (4.43) gives

$$\Pi(q) = \max(\Pi(q \mid p) * \Pi(p), \Pi(q \mid \neg p) * \Pi(\neg p)) \qquad (4.44)$$

$$\Pi(\neg q) = \max(\Pi(\neg q \mid p) * \Pi(p), \Pi(\neg q \mid \neg p) * \Pi(\neg p)) \qquad (4.45)$$

Equations (4.44) and (4.45) are a generalization of modus ponens because when $\Pi(q \mid p) = 1$, $\Pi(\neg q \mid p) = 0$ and $\Pi(p) = 1$, $\Pi(\neg p) = 0$, we can deduce that $\Pi(q) = 1$ and $\Pi(\neg q) = 0$. Equation (4.43) induces the inequality $\Pi(q) \geq \Pi(q \mid p) * \Pi(p)$, from which we can derive inference schemas of modus ponens and modus tollens, respectively:

$$
(X) \quad
\begin{array}{ll}
\Pi(q \mid p) & \geq a \\
\Pi(p) & \geq b \\
\hline
\Pi(q) & \geq a * b
\end{array}
\qquad
(XI) \quad
\begin{array}{ll}
\Pi(q \mid p) & \geq a \\
\Pi(p) & \leq b \\
\hline
\Pi(q) & \leq \sup\{s \in [0, 1], a * s \leq b\} \\
& \triangleq a * \rightarrow b
\end{array}
$$

These schemas can be combined into another, similar to (IX):

(XII)
$$\Pi(q \mid p) \geq a$$
$$\Pi(p \mid q) \geq a'$$
$$\Pi(p) \in [b, b']$$

$$\overline{\Pi(q) \in [a * b, a' * \to b']}$$

Equation (4.45) can be rewritten in terms of necessity measures by putting $N(q \mid p) = 1 - \Pi(\neg q \mid p)$, namely,

$$N(q) = \min(N(q \mid p) \perp N(\neg p), N(q \mid \neg p) \perp N(p)) \qquad (4.46)$$

where $s \perp t = 1 - (1 - s) * (1 - t)$. When $*$ is min, $\perp$ is max; more generally, $\perp$ is a disjunction operator satisfying the conditions

1. If $r \leq s$ and $t \leq u$, then $r \perp t \leq s \perp u$ (monotonicity)
2. $\forall s \in [0, 1]$, $s \perp 1 = 1 \perp s = 1$
3. $\forall s \in [0, 1]$, $0 \perp s = s \perp 0 = s$

Equation (4.46) induces the inequality $N(q) \geq \min(N(q \mid p), N(p))$, which yields the following schemas of modus ponens and modus tollens, respectively:

(XIII)
$$N(q \mid p) \geq a$$
$$N(p) \geq b$$

$$\overline{N(q) \geq \min(a, b)}$$

(XIV)
$$N(q \mid p) \geq a$$
$$N(q) \leq b$$

$$\overline{N(p) \leq 1 \text{ if } a \leq b}$$
$$b \text{ if } a > b$$

They can be recombined according to the schema for reasoning by equivalence:

(XV)
$$N(q \mid p) \geq a$$
$$N(p \mid q) \geq a'$$
$$N(p) \in [b, b']$$

$$\overline{N(q) \in \left[\min(a, b), \begin{cases} 1 & \text{if } a' \leq b' \\ b' & \text{if } a' > b' \end{cases} \right]}$$

Note that schemas (XIII)–(XV) do not depend on the operation $*$ which defines the conditioning. There are probabilistic inference schemas analogous to (XIII) and (XIV), based on conditional probability (Suppes [43]):

$$\text{(XVI)} \quad \frac{\begin{array}{ll} P(q \,|\, p) & \geq a \\ P(p) & \geq b \end{array}}{\begin{array}{ll} P(q) & \geq a \cdot b \end{array}} \qquad \text{(XVII)} \quad \frac{\begin{array}{ll} P(q \,|\, p) & \geq a \\ P(q) & \leq b \end{array}}{\begin{array}{ll} P(p) \leq 1 & \text{if } a = 0 \\ \min(1, b/a) & \text{if } a \neq 0 \end{array}}$$

Remark. $N(q \,|\, p)$ and $N(p \,|\, q)$, which appear together in schema (XV), can be interpreted, respectively, as degrees of "sufficiency" and of "necessity" for q to be true when p is true. This notion of attaching measures of sufficiency and necessity to a rule is the basis of the system of approximate reasoning adopted in PROSPECTOR [15]; the idea is developed therein in a probabilistic neo-Bayesian framework.   □

4.2.1.3. Synthesis

In both the logical and the functional approaches, the best intervals enclosing the degrees of necessity and possibility or of probability are always of the form $[a * b, a' * \to b']$, where $*$ is a conjunction operator and $* \to$ is an implication operator defined in terms of $*$ by (Prade [32], Weber [47], Dubois and Prade [58])

$$a * \to b = \sup\{s \,|\, a * s \leq b\} \tag{4.47}$$

This result holds for inference schemas based on material implication as well as for those based on conditioning. Probabilistic inference schemas give more precise results when the conditioning approach is used; in the case of possibilistic schemas this is not the case because schemas (IX), (XII), and (XV) give identical values of necessity, but possibility values which are not comparable [except when $a * b = \max(0, a + b - 1)$ in (XII), which yields an interval containing all others]. On the other hand, certain schemas (not the same ones on both approaches) yield no nontrivial deduction.

Implication-based and conditioning-based inference models are especially closely related when the triangular norm defining conditional possibility is taken as the minimum operation, as proposed by Hisdal [64]. In this case the least specific solution of (4.43), i.e., the maximal solution in the sense of fuzzy set inclusion [see (1.43)], is

$$\Pi(q \,|\, p) = 1 \qquad \text{if } \Pi(p) = \Pi(p \wedge q)$$
$$= \Pi(p \wedge q) \qquad \text{otherwise}$$

Hence $\Pi(q \mid p)$ is very closely related to $\Pi(p \wedge q) = 1 - N(p \rightarrow \neg q)$. More specifically, $N(q \mid p) = 1 - \Pi(\neg q \mid p)$ is very closely related to $N(p \rightarrow q)$, since

$$N(q \mid p) = \begin{cases} 0 & \text{if } N(\neg p) = N(p \rightarrow q) \\ N(p \rightarrow q) & \text{if } N(\neg p) < N(p \rightarrow q) \end{cases}$$

This result explains why the inference schemas (IX) and (XV) are identical [when $\Pi(p)$ is left unconstrained in (IX)]. It can then be proved that when $* = \min$ the inference schema expressed by (4.44) and (4.45), and based on conditional possibility, is completely consistent with (but stronger than) its counterpart based on necessity measures and logical implication, namely,

$$N(r) \geqslant \max(\min(N(p \rightarrow r), N(p)), \min(N(\neg p \rightarrow r), N(\neg p))),$$

$$r = q, \neg q$$

This inference schema subsumes (I) and (V), and can be written in matrix form, as well as (4.43). See Dubois and Prade [60] for a careful study of the links between $\Pi(q \mid p)$ and $N(p \rightarrow q)$. In particular, $\Pi(q \mid p)$ is a weight referring to the rule " if p then not q," as can be seen from the above results. A systematic application of (4.44) and (4.45) to automated approximate reasoning is given in Farreney and Prade [61]. In a probabilistic setting, implication and conditioning are not so closely related, as is shown in [60].

To sum up, it must be remembered that both approaches agree in that they lead to very similar calculations on degrees of uncertainty; this indicates that the calculations are well founded, even though this kind of investigation is still in its early stages. In practice, moreover, it is not always feasible to get an expert to be explicit about the mathematical nature (e.g., possibility, probability, necessity) of the weighting he proposes; any more than he may know whether the rule "if p then q" is to be understood as a material rather than a conditional implication. The fact that the results obtained are robust with regard to these two factors is reassuring in this respect.

N.B.1. A possible third approach would make direct use of multivalued-logic interpretations of the implication $p \rightarrow q$ to interpret the rule "if p then q." The results obtained are similar to those obtained here (see Prade [34, 36] for the details). This approach underlies the reasoning schema used in the inference system PROTIS [42]; this schema corresponds to (XII) in which $\Pi(q \mid p)$ is replaced by the truth value of the implication $p \to q$ [based on Lukasiewicz implication, i.e., using $a * b = \max(0, a + b - 1)$ and $a' * \to b' = \min(1, 1 - a' + b')$], and $\Pi(p)$ is replaced by the truth value of p. Nonetheless, interpretation of such truth values in $[0, 1]$ for nonvague propositions raises problems.   □

N.B.2. Shafer's measures of credibility and plausibility can be used instead of possibility, necessity, and probability measures in the inference schemas studied in this section. The reader is referred to Garvey *et al.* [63], Dubois and Prade [59], and Chatalic *et al.* [57].   □

4.2.2. Complex Premises

Let the proposition p be the conjunction of two more elementary propositions p_1 and p_2. If we suppose that the variables implicit in p_1 and in p_2, respectively, are distinct and noninteractive, then by (4.21) we can write

$$\Pi(p_1 \wedge p_2) = \min(\Pi(p_1), \Pi(p_2)) \tag{4.48}$$

Moreover, we always have

$$N(p_1 \wedge p_2) = \min(N(p_1), N(p_2)) \tag{4.49}$$

Analogous results hold if p is the disjunction of two propositions, from formulas (4.23) and (4.24). The analog of (4.48) and (4.49) are used, empirically and without prerequisite hypotheses, in the expert system MYCIN. When the variables are interactive, (4.48) is no longer valid; the interaction must be explicitly taken into account in the calculation.

4.2.3. Combining Degrees of Uncertainty Relative to the Same Proposition

4.2.3.1. Approach Based on Possibility Theory

Suppose that the degree of uncertainty for a proposition p is given in the form of the degree of possibility $\Pi_i(p)$, and the corresponding degree of necessity $N_i(p)$, with reference to a source i (which might be a rule like "if q_i then p"). When there are n sources, we may wish to combine the pairs $(\Pi_i(p), N_i(p))$ into a single index $(\Pi(p), N(p))$. Note that we always have

$$\max(\Pi_i(p), 1 - N_i(p)) = 1 \tag{4.50}$$

A natural idea is to extract the nonconflicting parts of the data from the different sources. We can view the data from source i as a fuzzy set on the reference set $\{p, \neg p\}$, F_i say, such that

$$\mu_{F_i}(p) = \Pi_i(p), \qquad \mu_{F_i}(\neg p) = 1 - N_i(p) = \Pi_i(\neg p)$$

This fuzzy set is always normalized because of (4.50). The fuzzy set corresponding to $(\Pi(p), N(p))$ can be defined by

$$F = \bigcap_{i=1}^{n} F_i$$

where the intersection is effected by the min operator. This amounts to putting

$$\Pi(p) = \min_i \Pi_i(p), \qquad N(p) = \max_i N_i(p)$$

Nevertheless we cannot be sure that the condition $\max(\Pi(p), 1 - N(p)) = 1$ still holds; it will not hold when there is conflict between the sources, i.e., when according to one source p is more possible than $\neg p$, while according to another the reverse is true. In the case of such conflict, one may decline to combine the data and start asking questions about the trustworthiness of the sources. If the sources are reliable and their data bear on the same problem, then we are led to normalize F by dividing its membership function by $\max(\Pi(p), 1 - N(p))$. Thus we obtain the formulas

$$\Pi(p) = \frac{\min_i \Pi_i(p)}{\max(\min_i \Pi_i(p), \min_i (1 - N_i(p)))} \tag{4.51}$$

$$N(p) = 1 - \frac{\min_i (1 - N_i(p))}{\max(\min_i \Pi_i(p), \min_i (1 - N_i(p)))} \tag{4.52}$$

An important problem is finding out whether the data from several sources support each other or not [which, prudently, is not presupposed in (4.51) and (4.52)]. This problem is especially encountered when one wishes to combine two rules like "if p_1 then q" and "if p_2 then q" in the form "if r then q" where r combines p_1 and p_2. For example, $\Pi(q \mid p_1 \wedge p_2)$ cannot be readily expressed in terms of $\Pi(q \mid p_1)$ and $\Pi(q \mid p_2)$ without introducing further assumptions: in some cases we may have

$$\Pi(q \mid p_1 \wedge p_2) \geq \max(\Pi(q \mid p_1), \Pi(q \mid p_2))$$

or

$$\Pi(q \,|\, p_1 \wedge p_2) \leqslant \min(\Pi(q \,|\, p_1), \Pi(q \,|\, p_2))$$

It is, in particular, necessary to know whether the rule "if p_1 then q" can be asserted independently of the truth of p_2. This problem, which is related to the noninteraction problem, is among the open questions of approximate reasoning. It is extensively discussed in Ref. 69.

4.2.3.2. Approach Based on Dempster's Rule

Let it now be supposed that the data are given in the form of plausibility and credibility measures on the set $\mathscr{P}$ of propositions (see Section 1.1). Each source i is supposed to provide a probabilistic weighting m_i in the sense of equation (4.2). Dempster's rule [6], which generalizes Bayes's theorem (see also Shafer [39]) allows m_1 and m_2 to be combined into a new weighting m_{12} which satisfies (4.2) and is defined as follows:

- $\forall p \in \mathscr{P}$, calculate $m(p) = \sum \{m_1(q) \cdot m_2(r) \,|\, p = q \wedge r\}$ (4.53)

- Then $\forall p \neq \mathbb{0}$ put $m_{12}(p) = \dfrac{m(p)}{1 - m(\mathbb{0})}$ (4.54)

Dempster's rule is associative and commutative, which means that we can study the case of two sources without loss of generality. If m_1 defines a probability measure on $\mathscr{P}$, then m_{12} also defines a probability measure. If in addition m_2 is concentrated on p $(m_2(p) = 1)$ then we recover Bayes' formula for conditional probabilities.

Note that (4.54) amounts to normalizing the weighting m obtained from (4.53), which, in the case of very discordant sources, is dubious see Refs. 54 or 59). The quantity $m(\mathbb{0})$ measures the degree of conflict between the two sources; the effect of normalization is to mask this conflict, as in (4.51) and (4.52).

Let us, as previously, suppose that $\mathscr{P} = \{\mathbb{0}, p, \neg p, \mathbb{1}\}$. Let source i give credibility $\mathrm{Cr}_i(p) = b_i$ and plausibility $\mathrm{Pl}_i(p) = a_i$, where $a_i \geqslant b_i$ [cf. (4.8)]. From (4.3) and (4.5) it is easy to verify that the corresponding weighting m_i satisfies

$$m_i(p) = b_i, \qquad m_i(\neg p) = 1 - a_i, \qquad m_i(\mathbb{1}) = a_i - b_i$$

where $m(\mathbb{1})$ is the weight given to complete ignorance. Pl_i is a possibility measure if $a_i = 1$ (emphasizing p) or $b_i = 0$ (emphasizing $\neg p$). It is a probability measure if $a_i = b_i$. Dempster's rule now gives the following results obtained for the two pairs of coefficients (a_1, b_1) and (a_2, b_2):

$$m(\emptyset) = b_1(1 - a_2) + b_2(1 - a_1)$$
$$m(p) = a_1b_2 + a_2b_1 - b_1b_2$$
$$m(\neg p) = 1 - b_1 - b_2 + b_1a_2 + b_2a_1 - a_1a_2 \tag{4.55}$$
$$m(\mathbb{1}) = (a_1 - b_1)(a_2 - b_2)$$

These formulas enable us to derive various rules of combination which have been introduced in the literature without reference to the Shafer–Dempster approach:

- m_1 and m_2 *define probability measures.* In this case, $a_1 = b_1 = P_1(p)$, $a_2 = b_2 = P_2(p)$ and we find that $m_{12}(\mathbb{1}) = 0$ (so that m_{12} defines a probability measure), and

$$m_{12}(p) = P_{12}(p) = \frac{a_1a_2}{1 - a_1 - a_2 + 2a_1a_2} = \sigma_0(a_1, a_2) \tag{4.56}$$

  We thus find Silvert's symmetric sum σ_0 (see Chapter 3, Section 3.1.2.3.). This combination formula is employed by Kayser [23].
- m_1 and m_2 *define nonconflicting possibility measures.* $[m_1(p) > 0$ and $m_2(p) > 0]$. We then have $a_1 = a_2 = 1$, $N_i(p) = b_i$, $i = 1, 2$, and we find

$$m_{12}(p) = b_1 + b_2 - b_1b_2 = N_{12}(p)$$
$$m_{12}(\mathbb{1}) = (1 - b_1)(1 - b_2) \tag{4.57}$$

(4.57) is used in MYCIN [40] in order to combine, on one hand, the belief measures $MB(h, e_1)$ and $MB(h, e_2)$, and on the other hand the disbelief measures $MD(h, e_1)$ and $MD(h, e_2)$ (see the Remark in Section 4.1.4). Clearly the credibility measure thus obtained is a necessity measure. Combination by Dempster's rule can therefore easily be compared with (4.51)–(4.52), which would give

$$N_{12}(p) = \max(b_1, b_2) = m_{12}(p)$$

and

$$m_{12}(\mathbb{1}) = \min(1 - b_1, 1 - b_2)$$

That is to say that Dempster's rule engenders mutual reinforcement of the data derived from different sources.

Equations (4.55) suggest other special cases such as the following:

- A possibility measure derived from source 1, and a probability measure from source 2 ($a_1 = 1$, $b_2 = a_2$). The resulting combination is a probability measure.
- Two conflicting possibility measures ($a_1 = 1$, $b_2 = 0$). We then obtain a confidence measure which is neither a probability nor a possibility. It can be verified that

$$\text{Pl}_{12}(p) = \frac{a_2}{1 - b_1(1 - a_2)}, \qquad \text{Cr}_{12}(p) = \frac{a_2 b_1}{1 - b_1(1 - a_2)} \qquad (4.58)$$

while (4.51) and (4.52) give

$$\Pi_{12}(p) = \frac{a_2}{1 - \min(b_1, 1 - a_2)}, \qquad N_{12}(p) = \frac{\max(0, a_2 + b_1 - 1)}{1 - \min(b_1, 1 - a_2)} \qquad (4.59)$$

Note the evident structural similarity between (4.58) and (4.59), the difference between them arising from the triangular norms which have been used. (4.59) is the formula used in MYCIN [40], up to the analogy made in Section 4.1.

Many problems can arise in combining uncertain items of information. The results given by Dempster's rule are extremely sensitive to the numerical values of the inputs, especially with conflicting sources. Consider in fact an example where there are just three mutually exclusive alternatives a, b, c and we apply (4.53) and (4.54) to the two basic probability assignments

$$m_1(a) = 0.0, \qquad m_1(b) = 0.1, \qquad m_1(c) = 0.9$$
$$m_2(a) = 0.9, \qquad m_2(b) = 0.1, \qquad m_2(c) = 0.0$$

corresponding to two extremely dissonant sources of information; we find the sharp result

$$m(a) = 0.0, \qquad m(b) = 1.0, \qquad m(c) = 0.0$$

as pointed out by Zadeh [54]. However, if m_1 and m_2 are slightly altered to

$$m_1'(a) = 0.01, \quad m_1'(b) = 0.1, \quad m_1'(c) = 0.89$$
$$m_2'(a) = 0.89, \quad m_2'(b) = 0.1, \quad m_2'(c) = 0.01$$

then (4.53) and (4.54) give

$$m'(a) \simeq 0.32, \quad m'(b) \simeq 0.36, \quad m'(c) \simeq 0.32$$

which is a very unsharp result. The result of the first case may be justified by interpreting zero as *complete* certainty of the impossibility of alternatives a and c, respectively, whereupon, however improbable b may be, it remains the only possible alternative (Shafer [39]). This reasoning can be questioned if zero may also represent a very small value of probability, in which case the alternatives a and c are not definitely ruled out by the first and second sources, respectively. But then the result may (as above) change drastically from a strong degree of certainty in favor of b (when 0.0 is used) to a very uncertain outcome (as when 0.01 is used). This rather counterintuitive behavior of Dempster's rule may create anomalies when blindly applied in expert systems where degrees of uncertainty are subjectively assessed. See [59] for more examples and discussion of this point. Note that this sensitivity problem is less acute when the "possibilistic" schema (4.51) and (4.52) is used, although the question of the "meaning of zero" is still outstanding in this setting.

A further difficulty is linked to the interpretation of uncertain rules. For instance, if we know with complete certainty that p is true and that the rule "if p (true) then q (true)" holds, and we know with certainty equal to $\lambda < 1$ that the rule "if q (true) then r (true)" holds, then we can deduce that r is true with a certainty at least equal to λ. This can be derived in a probabilistic as well as in a possibility/necessity-based approach, and using models based on logical implication as well as models based on conditioning. On the other hand, as Zadeh [53] points out, from the two rules "All A's are B's" and "QB's are C's," it *cannot* be deduced that "QA's are C's" where Q is a numerical or linguistic quantifier (e.g., 80%, most) which is different from "all" (i.e., the universal quantifier $\forall$). At first glance the behaviors of these two schemas may seem contradictory; but in fact they are not. Indeed, if we know with complete certainty that x is an A, we deduce with complete certainty that x is a B from the first rule; and, from the second (*in the absence of other information*), that x is a C with a degree of certainty that is the greater, the closer Q is to "All." This corresponds to the first schema.

Finally, blind combination does not look reasonable when dealing with the results of rules having varying degrees of generality. The fact that an uncertain conclusion may be no longer acceptable (in the face of further information not used in its derivation) makes the use of an "intelligent" combination procedure necessary in combining uncertain items of information (especially in the case of conflict). For instance, if we should also have the rule "$Q'A$'s are C's" in the above example, the conclusion obtained directly by this rule for an x which is A should be preferred to the conclusion obtained by using the other two rules in cascade (indeed the exceptions to the rule "QB's are C's" may be mainly among the x's which are both A and B); see [61] for further discussions.

4.3. Inference from Vague or Fuzzy Premises

In this section we shall suppose that the propositions to be handled are of the form $p = $ "X is A," where A is a fuzzy set on the reference set S. In this case the lattice of propositions being considered is not Boolean, and we shall introduce explicit interpretations of the propositions, since two distinct propositions p and p' can be much closer to each other than when they are classical propositions. In particular, there will be an extended form of modus ponens, according to which we can sometimes deduce a nontrivial conclusion $q' \neq q$ from the rule "if p then q" and a proposition $p' \neq p$.

4.3.1. Representation of the Rule "if X is A, then Y is B"

Suppose the proposition p expresses a restriction on the values of a variable X, and the proposition q a restriction on the possible values of a variable Y. A causal link from X to Y can be represented by a conditional possibility distribution $\pi_{Y|X}$ (or probability distribution $p_{Y|x}$) which constrains the values of Y for a given value of X. Given a possibility distribution representing p, we can calculate the possibility distribution π_Y restricting Y from the formula

$$\forall t \in T, \qquad \pi_Y(t) = \sup_{s \in S} \pi_{Y|X}(t, s) * \pi_X(s) \qquad (4.60)$$

where $*$ is a triangular norm; (4.60) is the possibilistic analog of the formula

$$\forall t \in T, \qquad p_Y(t) = \sum_{s \in S} p_{Y|X}(t, s) \cdot p_X(s) \qquad (4.61)$$

in probability theory.

Note that $\pi_{Y|X}(t, s) = \Pi(q \mid p)$ if $q =$ "$Y = t$" and $p =$ "$X = s$," and π_Y is the projection, onto the reference set T of Y, of the possibility distribution $\pi_{(X,Y)} = \pi_{Y|X} * \pi_X$ defined according to (4.43).

In this context, the rule "if X is A then Y is B" will be represented by the inequality, derived from (4.42) (see Refs. 32 and 12)

$$\forall t \in T, \qquad \mu_B(t) \geqslant \sup_{s \in S} \pi_{Y|X}(t, s) * \mu_A(s) \qquad (4.62)$$

where μ_A and μ_B are the possibility distributions associated with X and Y, respectively, $\pi_{Y|X}$ being unknown; the inequality derives from the fact that "Y is B'" can be deduced from "Y is B" as soon as B' corresponds to a possibility distribution which contains the one associated with B. The maximal solution of (4.62) for a continuous triangular norm is given (see Refs. 70 and 12) by

$$\pi_{Y|X}(t, s) = \mu_A(s) *\!\rightarrow \mu_B(t) \qquad (4.63)$$

where $*\!\rightarrow$ is defined by (4.47). In particular,

$$a * b = \min(a, b) \Rightarrow a *\!\rightarrow b$$

$$= \begin{cases} 1 & \text{if } a \leqslant b \\ b & \text{if } a > b \end{cases} \qquad \text{(Gödel's implication [38])}$$

$$a * b = a \cdot b \Rightarrow a *\!\rightarrow b$$

$$= \begin{cases} 1 & \text{if } a = 0 \\ \min(1, b/a) & \text{otherwise} \end{cases} \qquad \text{(Goguen's implication [19])}$$

$$a * b = \max(0, a + b - 1) \Rightarrow a *\!\rightarrow b$$

$$= \min(1, 1 - a + b) \qquad \text{(Lukasiewicz's implication [37])}$$

For other possible operations and their interrelationships, see [12].

4.3.2. "Generalized" Modus Ponens

Now we are in a position to consider Zadeh's "generalized modus ponens" [51].

	If X is A	then	Y is B
(XVIII)	X is A'		

$$\text{then} \quad Y \text{ is } B'$$

where $\mu_{B'} = \pi_Y$ can easily be calculated from (4.60) and (4.63), which gives

$$\forall\, t \in T, \qquad \mu_{B'}(t) = \sup_{s \in S} (\mu_A(s) *\!\to \mu_B(t)) * \mu_{A'}(s) \qquad (4.64)$$

which we shall write symbolically as $B' = A' \circ (A *\!\to B)$. We can then verify [32, 12] that $A \circ (A *\!\to B) = B$ and that

- $\forall$ normalized A', $A' \subseteq A \Rightarrow B' = B$
- $\forall\, A'$, $\qquad B' \supseteq B$
- $\forall\, A'$, $\qquad \mu_{B'}(t) \geq \sup_s \{\mu_{A'}(s) \mid \mu_A(s) = 0\} = \mu_{\mathrm{CP}(A;A')}(0) \qquad (4.65)$

The inequality (4.65) means that a uniform degree of indetermination appears whenever a significant part of A' is not included in A, which seems intuitively natural. In particular, if $\exists\, s \in S$, $\mu_{A'}(s) = 1$ and $\mu_A(s) = 0$ then $B' = T$ (complete indeterminacy). Therefore, if we know that A' is "close to" (but sensibly different from) A in the sense of a certain metric, then the generalized modus ponens (XVIII) is not an adequate model to deduce from "X is A'" and from "if X is A then Y is B" that "Y is B'" where B' is "close to" B; this can only be done if we have further information on the continuity and monotonicity of the causal relation $X \to Y$ in the neighbourhood of (A, B) (see [11]). For example, from a rule like "if a tomato is red then it is ripe" and from the fact "this tomato is very red," generalized modus ponens does not allow us to draw the conclusion "this tomato is very ripe," which we can only do if we know that the degree of ripeness increases with the degree of redness.

Moreover, we can verify that from fuzzy rules of the form "if X is A then Y is B" and "if Y is B' then Z is C," we can construct the fuzzy rule "if X is A then Z is C" whenever $B \subseteq B'$ (see [12]), which leads directly by generalized modus ponens to the same result for Z as the application of the two rules in series.

Note that some authors (Mamdani [25]) have put $B' = A' \circ (A \times B)$, where $A \times B$ is the Cartesian product of A and B. $A \times B$ is a solution of (4.62) for $* = \min$, but not the maximal one, which gives a degree of arbitrariness to the corresponding result B'.

The inference scheme (XVIII) can be formulated in terms of truth values, by means of the notion of the compatibility of two fuzzy propositions introduced in Section 4.1.4, in the form (Baldwin [2])

$$\mu_{\mathrm{CP}(B;B')}(v) = \sup_{u \in [0,\,1]} (u *\!\to v) * \mu_{\mathrm{CP}(A;A')}(u) \qquad (4.66)$$

$$\mu_{B'}(t) = \mu_{CP(B;B')}(\mu_B(t)) \tag{4.67}$$

(4.66) and (4.67) are equivalent to (4.64).

More generally, if the premise "if X is A then Y is B" is weighted by a fuzzy truth value τ represented by a characteristic function μ_τ on the interval $[0, 1]$, which can be interpreted as

$$\tau = CP(A \ast\!\!\rightarrow B; X \rightarrow Y) \tag{4.68}$$

with $\mu_{A \ast\rightarrow B} = \mu_A \ast\!\!\rightarrow \mu_B$ and $\mu_{X \rightarrow Y} = \pi_{Y|X}$, then we finally obtain

$$\pi_{Y|X}(t, s) = \mu_\tau(\mu_A(s) \ast\!\!\rightarrow \mu_B(t)) \tag{4.69}$$

and

$$\forall t \in T, \qquad \mu_{B'}(t) = \sup_{s \in S} \mu_\tau(\mu_A(s) \ast\!\!\rightarrow \mu_B(t)) \ast \mu_{A'}(s) \tag{4.70}$$

N.B. This transformation clearly has no meaning when $A \ast\!\!\rightarrow B$ is a classical set and τ is a truth value expressing neither truth nor falsehood $[\mu_\tau(0) = \mu_\tau(1) = 0]$ (see [32]). □

Remark. We can similarly handle modus tollens with vague premises; generalized modus tollens is represented by the equation [12]:

$$\forall s \in S, \qquad \mu_{A'}(s) = \sup_t (\mu_A(s) \ast\!\!\rightarrow \mu_B(t)) \ast \mu_{B'}(t) \tag{4.71}$$

provided that $\ast\!\!\rightarrow$ is its own contrapositive, i.e.,

$$x \ast\!\!\rightarrow y = (1 - y) \ast\!\!\rightarrow (1 - x)$$

in order to ensure equivalence between the rules "if X is A then Y is B" and "if Y is non-B then X is non-A," i.e.,

$$A \ast\!\!\rightarrow B = \bar{B} \ast\!\!\rightarrow \bar{A}$$

We thus verify, for instance, taking

$$x \ast y = \max(0, x + y + 1), \qquad x \ast\!\!\rightarrow y = \min(1, 1 - x + y)$$

that $A' = \bar{A}$ when $B' = \bar{B}$ (with $\mu_{\bar{A}} = 1 - \mu_A$, $\mu_{\bar{B}} = 1 - \mu_B$). On the other hand, when $* = \min$, $* \rightarrow$ is not its own contrapositive, and it is therefore the latter that must be used to define modus tollens [13].   □

4.3.3. Complex Premises

Complex fuzzy propositions of the form "X_1 is A_1 and X_2 is A_2" can be naturally represented by Cartesian products of fuzzy sets (cf. Section 4.1.4) when the variables X_1 and X_2 are noninteracting. We thus obtain "(X_1, X_2) is $A_1 \times A_2$." Similarly a complex fuzzy proposition "X_1 is A_1 or X_2 is A_2" can be represented by means of the Cartesian coproduct $+$, as "(X_1, X_2) is $A_1 + A_2$." The presence of interaction, or indeed compensation, between the variables must be allowed for explicitly, or implicitly as in the approach of Chapter 3 to the combination of fuzzy sets. Generalized modus ponens can be trivially extended to complex premises by following the analysis of Section 3.2 using conditional possibility distributions of the form $\Pi_{Y \mid (X_1, X_2)}$ and so on.

4.3.4. Combining Possibility Distributions

Suppose one source of information asserts that "X is A_1", while another asserts that "X is A_2", each assertion being represented by a possibility distribution $\pi_i = \mu_{A_i}$, $i = 1, 2$.

One way to combine these assertions would be to take the intersection of their possibility distributions and normalize the result. Let

$$\forall s \in S, \qquad \pi_{12}(s) = \frac{\pi_1(s) * \pi_2(s)}{\sup_{s \in S} \pi_1(s) * \pi_2(s)} \qquad (4.72)$$

where $*$ is a fuzzy set-theoretic intersection (Chapter 3, Section 3.1.2). Equation (4.72) is in the same spirit as formulas (4.51) and (4.52) which emerge as special cases on taking $* = \min$ and $S = \{p, \neg p\}$. This technique presupposes that the two sources are reliable, so that the information can be sharpened by "cross-validation," where the reliability of the sources will justify the normalization of the resulting possibility distribution. To choose $* \neq \min$ means that coherent data are mutually reinforcing, in that the result will be more specific (in the sense of Chapter 1, Section 1.6.2.2). For example, if $A_1 = A_2 = A$, $\pi_{12} = \mu_A * \mu_A < \mu_A$.

When the two sources are in severe conflict, their reliability becomes dubious; one may then decide to adopt the union of the respective corresponding fuzzy sets, say

$$\forall s \in S, \qquad \pi_{12}(s) = \pi_1(s) \perp \pi_2(s) \tag{4.73}$$

where $\perp$ is a fuzzy set-theoretic union (Chapter 3, Section 3.1.2), usually max. Note that π_{12} will be normalized if π_1 and π_2 are. However, (4.73) can bring about a marked reduction of precision of the data.

Observe that Dempster's rule [cf. (4.53) and (4.54)], using weightings m_1 and m_2 on π_1 and π_2 [constructed by inverting formula (1.29)] to reconcile the data from the two sources, will not give a weighting m_{12} corresponding to any possibility distribution (see [10]).

Another situation giving rise to combination of possibilistic data arises when n rules "if X is A_i then Y is B_j" $(i = 1, n)$ are given to describe a causal link from X to Y. Each rule i is represented by a fuzzy set $A_i \ast\!\!\to B_i$ [cf. (4.63)], and the combined collection of rules can be represented by the intersection $\bigcap_i (A_i \ast\!\!\to B_i)$, where $\bigcap$ is realized by the "min" operator. This result is proved in [12]. Generalized modus ponens for several rules can then be expressed as

$$\forall t, \qquad \mu_{B'}(t) = \sup_s \left\{ \min_{i=1,n} (\mu_{A_i}(s) \ast\!\!\to \mu_{B_i}(t)) \ast \mu_{A'}(s) \right\} \tag{4.74}$$

If generalized modus ponens is applied to each rule separately and the intersection of the results is taken, it can be shown that the possibility distribution so obtained may be much less precise than that obtained from (4.74). A method for implementiing (4.74) is given in Ref. 26.

4.4. Brief Summary of Current Work and Systems

Until recently, there were few systems for approximate reasoning based on possibility theory, and most of the existing ones only use some of the results of this chapter. Developing an idea of Zadeh [50], a form of generalized modus ponens has been used for the first time to handle fuzzy rules by Mamdani [25], who has applied it to the control of continuous industrial processes. See Ref. 71 for an extensive account of recent applications of this methodology. Other forms of generalized modus ponens have been used in medical diagnosis by Lesmo, Saitta, and Torasso [24], Adlassnig [1], Buisson et al. [72], and Buckley et al. [56], and in systems for management decision support by Whalen and Schott [48] and Ernst [16]. These various forms of generalized modus ponens differ according to the choice of the operator $\ast\!\!\to$ in (4.64), made independently of the choice of the operator $\ast$ which is usually the min operator. Mizumoto and others [29, 30] have compared numerous forms of modus ponens, generalized with various operators. See also Baldwin and Pilsworth [3], Yager [49].

Schemas for reasoning with uncertain rules have been implemented by Soula *et al.* [41, 42] (see also Fieschi [67]) in the medical expert systems SPHINX and PROTIS, and by Tong *et al.* [44] in an information retrieval system; these schemas are based on multivalued logics, as is ARIES, a general inference engine [55]. Detachment rules for fuzzy truth values in multivalued logics have been studied in [7]; a similar approach has been applied by Tsukamoto [46] to problems of process control; in terms of computation this particular technique turns out to be equivalent to generalized modus ponens in the form (4.66) and (4.67); see [13]. Studies that consider rules that are both fuzzy and uncertain are still very few. Ishizuka, Fu, and Yao [22] study imprecise and uncertain data in a rule-based expert system in civil engineering (SPERIL [21]); these authors evaluate uncertainty in terms of Shafer credibility functions applied to fuzzy events (Ishizuha [20]). Martin-Clouaire and Prade [27, 28] propose a unified treatment of imprecision and uncertainty within the framework of possibility theory with a view to application to expert systems in geology; in the following section, their approach is applied to a small-scale example of evaluation of applications for a job. The corresponding (nondedicated) inference engine is given in Appendix B, and is called SPII [65]. Another inference engine, called TAIGER [62], is currently under development in the context of financial analysis.

Among general tools which can be used to develop expert systems incorporating fuzzy rules, may be mentioned fuzzy pattern-matching [5], which gives the possibility $\Pi(A; A')$ and the necessity $N(A; A')$ that a possibly fuzzy condition "X is A" should hold, given the (possibly also fuzzy) data "X is A'" on the current situation. The quantities $\Pi(A; A')$ and $N(A; A')$, which approximate the compatibility function $CP(A; A')$ [cf. (4.31) and (4.32)] that appears in generalized modus ponens in the form (4.66) and (4.67), can be used to evaluate the coefficients b and b' in the schemas for approximate reasoning of Section 4.2, when A is nonfuzzy.

For more details on fuzzy expert systems and other kinds of information systems, the reader may consult the edited volumes Refs. 66 and 73.

4.5. Example

The fundamental techniques described in this chapter have been implemented in a simple inference engine (backward chaining over propositions) for which a complete LE-LISP program is given in an appendix (implemented on a VAX 750 under VMS). In this section we give a simple example of its application, and the essential results concerning an efficient procedure for effecting a generalized modus ponens [taking $* = $ min in (4.47) and (4.64)].

The illustrative example concerns evaluation of candidates for a job. It is based on the following

- The results (not necessarily available) of various tests (e.g., intelligence, French) in which the candidate may be marked imprecisely (i.e., qualitatively).
- Criteria evaluated by consulting a dossier (e.g., adequacy of previous education) and by interview (e.g., adaptibility to change of work, main interests, special skills).

The following conventions are adopted in the program. A fact is an atom (e.g., f_3) whose value is a triple (attribute, object, value), where the *value* component may be precise or imprecise (fuzzy). When imprecise, this component is an atom (and its name then starts with "$\sim$") whose value is a possibility distribution which is continuous and trapezoidal (on $\mathbb{R}$, and represented by a quadruple). Uncertainty about a precise fact is associated with the property "cert" in memory; this uncertainty is represented by two numbers b, b' such that $b \leq b'$ and $b = 0$ or $b' = 1$, interpreted in accordance with (4.13) as degrees of necessity and of possibility that the fact should be true. A rule is an atom (e.g., r_7) whose value is a list consisting of two sublists: the first is a list of premise-triples connected by a conjunction, and the second contains conclusion-triples. In our example, each rule gives only one conclusion. In the case of an uncertain rule (i.e., a rule giving a precise but uncertain conclusion), the triple is followed by two coefficients a and a', respectively, interpreted as degrees of "sufficiency" and of "necessity" that the premises should be satisfied in order that the conclusion should hold. These two coefficients are lower bounds of $N(p \rightarrow q)$ and $N(q \rightarrow p)$, respectively, quantities that were introduced in Section 4.2.1.1. The inference mechanism is based on a reasoning schema that combines (I) and (V), namely, a particular case of (IX):

$$N(p \rightarrow q) \geq a$$
$$N(q \rightarrow p) \geq a'$$

(XIX) $\quad N(p) \geq b; \Pi(p) \leq b'$ with $\max(1 - b, b') = 1$

$$\overline{N(q) \geq \min(a, b), \qquad \Pi(q) \leq \max(1 - a', b')}$$

This schema can also be justified in terms of multivalued logic [3, 4, 35] used as proposed in Ref. 27. Note that if $\max(1 - b, b') = 1$ then we always have

$$\max(1 - \min(a, b), \max(1 - a', b')) = 1, \qquad \forall\, a, \forall\, a'$$

N.B. In the LISP program the operations $\min(a, b)$ and $\max(1 - a', b')$ can, respectively, be weakened to $a \cdot b$ and $1 - a' + a' \cdot b'$, or to $\max(0, a + b - 1)$ and $\min(1, 1 - a' + b')$; these options only seem to be valid in terms of multivalued logic; nonetheless the results obtained do not seem to be false in any case since they embrace the interval $[\min(a, b), \max(1 - a', b')]$. □

In our example, some uncertain rules have fuzzy premises.

The imprecise rules are those where the value-components of the premises and of the conclusion are possibility distributions. The LISP program in Appendix B presupposes that the distributions given in the rules are continuous and trapezoidal.

The rules provided enable simulation of reasoning represented by the following tree structure, which describes the set of possible chainings of the rules:

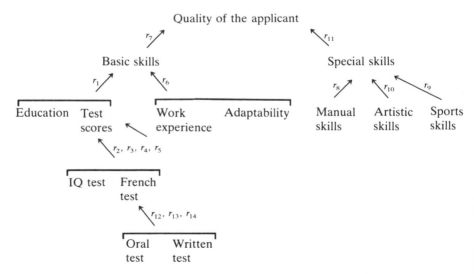

N.B. This example is of a very similar kind to the example of

Chapter 3 concerning the choice of a car, but the manner of representing the aggregation process which leads to the evaluation is radically different. □

The rules whose premises and conclusion are imprecise are treated by generalized modus ponens [cf. (4.64) with $* = \min$]; the collection r_2, r_3, r_4, r_5 is an example of such rules. The rules where possibility distributions do not occur, but which are uncertain, are treated by the schema (XIX) where b and b' are obtained as

$$b = \min_{i=1,n} b_i, \qquad b' = \min_{i=1,n} b_i'$$

where n is the number of premises and $[b_i, b_i']$ is the certainty interval of the fact which is associated with the ith premise and which belongs to the initial knowledge base or which results from a preliminary inference. For uncertain rules where some premises involve possibility distributions, b_i and b_i' are, for uncertain premises, calculated as the necessity and possibility, respectively, that the premise should hold [formulas (4.17), (4.18)]. These formulas do not always give results for which $\max(1 - b_i, b_i') = 1$, in which case the normalization

$$b_i^* = 1 - \frac{1 - b_i}{\max(1 - b_i, b_i')}, \qquad b_i'^* = \frac{b_i'}{\max(1 - b_i, b_i')}$$

agreeing with (4.51) and (4.52), always leads to a pair $(b_i^*, b_i'^*)$ which satisfies the constraint and which, therefore, is easier to interpret. This normalization emphasizes whether the premise is mainly satisfied $(b_i'^* = 1)$ or mainly not satisfied $(b_i^* = 0)$.

Last, the results from several uncertain rules are combined according to formulas (4.51) and (4.52).

Appendix B gives the LISP program which implements the example and the above procedures. It consists of the following:

- The inference engine, which handles rules that are uncertain or fuzzy or both.
- The initializer file, which defines the program's initialization function.
- The knowledge base, which contains the rules r_1–r_{14}, the definitions of fuzzy terms, and eight known facts for a candidate. Note

that the system does not necessarily require explicit information about the nine aspects that are at the termini of the decision tree.
- An example of a session in which the quality of the candidate described by the eight facts is evaluated, and the system furnishes on demand the intermediate evaluations.

Finally, the section "treatment of uncertain rules" of the inference engine is rewritten as a BASIC program given at the end of Appendix B, with a set of rules and a knowledge base of facts for a different example which the reader can easily reconstruct.

Appendix A

The program uses an efficient implementation of generalized modus ponens based on Gödel's multivalued implication and the use of "min." This program assumes that continuous possibility distributions are trapezoidal. We give below the essential results to help the reader understand the treatment of these distributions in the program.

Given the rule "if X is A then Y is B" and the fact "X is A'" we seek to compute B' such that the fact "Y is B'" is deduced. $\mu_{B'}$ is given by (4.64), where $* = \min$, i.e.,

$$\forall t, \qquad \mu_{B'}(t) = \sup_{s \in S} \min(\mu_{A'}(s), \mu_A(s) \to \mu_B(t)) \qquad (4.75)$$

where $\to$ is defined by

$$a \to b = 1 \qquad \text{if } a \leq b$$
$$= b \qquad \text{if } a > b$$

Letting $\lambda = \mu_B(t)$, the calculation of (4.75) is split into two steps:

1. Calculate the mapping of $[0, 1] \to [0, 1]$ expressed by

$$\forall \lambda, \qquad f(\lambda) = \sup_{s \in S} \min(\mu_{A'}(s), \mu_A(s) \to \lambda)$$

2. Calculate $\mu_{B'}$ as

$$\forall t, \qquad \mu_{B'}(t) = f(\mu_B(t))$$

Let us focus on step 1. Using the definition of $\rightarrow$ leads to

$$f(\lambda) = \max\left(\sup_{s \notin A_{\bar{\lambda}}} \mu_{A'}(s), \sup_{s \in A_{\bar{\lambda}}} \min(\lambda, \mu_{A'}(s))\right)$$

$$= \max(\Pi'(\overline{A_{\bar{\lambda}}}), \min(\lambda, \Pi'(A_{\bar{\lambda}}))) \tag{4.76}$$

where $A_{\bar{\lambda}}$ is the strong λ-cut of A (see Chapter 1, Section 1.4), and Π' is the possibility measure based on the distribution $\mu_{A'}$.

Note that

$$\text{if } \Pi'(\overline{A_{\bar{\lambda}}}) \geq \lambda \quad \text{then } f(\lambda) = \Pi'(\overline{A_{\bar{\lambda}}}) \geq \lambda$$

$$\text{if } \Pi'(\overline{A_{\bar{\lambda}}}) < \lambda \quad \text{then } \Pi'(A_{\bar{\lambda}}) = 1 \quad \text{and} \quad f(\lambda) = \lambda$$

Hence (4.76) can be simplified, in the general case, to

$$f(\lambda) = \max(\Pi'(\overline{A_{\bar{\lambda}}}), \lambda) \tag{4.77}$$

N.B. This result shows, if needed, that $B' \supseteq B$, $\forall A'$, as mentioned in Section 4.3.2. $\qquad\qquad\qquad\qquad\qquad\qquad\qquad\qquad\qquad\square$

Now let us suppose that A and A' are trapezoidal fuzzy intervals on a compact scale S. Denoting $A_{\bar{\lambda}} =]\underline{a}_\lambda, \bar{a}_\lambda[$, it is not difficult to work out that

$$\Pi'(\overline{A_{\bar{\lambda}}}) = \max(\mu_{[A', +\infty)}(\underline{a}_\lambda), \mu_{(-\infty, A']}(\bar{a}_\lambda))$$

where $\mu_{[A', +\infty)}$ and $\mu_{(-\infty, A']}$ are the fuzzy intervals defined in Chapter 3, Section 3.2. Introducing the auxiliary functions f^+ and f^- such that

$$f^+(\lambda) = \max(\mu_{[A', +\infty)}(\underline{a}_\lambda), \lambda) \tag{4.78}$$

$$f^-(\lambda) = \max(\mu_{(-\infty, A']}(\bar{a}_\lambda), \lambda)$$

we have $f = \max(f^+, f^-)$. Let us calculate f^+.

Following Chapter 2, Section 2.3.1, let $A = (\underline{a}, \bar{a}, \alpha, \beta)_{LL}$, $A' = (\underline{a}', \bar{a}', \alpha', \beta')_{LL}$ with $L(x) = \max(0, 1 - x)$.

The shape of f^+ depends on how $(\underline{a}, \underline{a} - \alpha)$ is situated with respect to $(\underline{a}', \underline{a}' - \alpha')$:

Case 1: $\underline{a}' \leq \underline{a} - \alpha$. See Figure 4.1. Clearly $f^+(\lambda) = 1$, $\forall \lambda \in [0, 1]$.

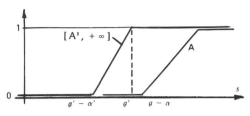

Figure 4.1. Case 1.

Case 2: $\underline{a}' - \alpha' \leqslant \underline{a} - \alpha < \underline{a}' \leqslant \underline{a}$. See Figure 4.2. Let $\lambda_0^+ = \mu_{[A',+\infty)}(\underline{a} - \alpha)$, $\lambda_1^+ = \mu_{[A,+\infty)}(\underline{a}')$; then

$$f^+(\lambda) = 1 \qquad \text{if } \lambda \geqslant \lambda_1^+$$
$$= \lambda_0^+ + \lambda(1 - \lambda_0^+)/\lambda_1^+ \qquad \text{otherwise}$$

Case 3: $\underline{a} - \alpha < \underline{a}' - \alpha'$, $\underline{a}' < \underline{a}$. See Figure 4.3. Let $\bar{\lambda}^+$ be such that $\mu_A(s) = \mu_{A'}(s) = \bar{\lambda}^+ < 1$; then

$$f^+(\lambda) = 1 \qquad \text{if } \lambda \geqslant \lambda_1^+$$
$$= \bar{\lambda}^+ + \frac{(\lambda - \bar{\lambda}^+)}{(\lambda_1^+ - \bar{\lambda}^+)}(1 - \bar{\lambda}^+) \qquad \text{if } \bar{\lambda}^+ < \lambda \leqslant \lambda_1^+$$
$$= \lambda \qquad \text{if } \lambda \leqslant \bar{\lambda}^+$$

Case 4: $\underline{a}' - \alpha' < \underline{a} - \alpha$, $\underline{a}' > \underline{a}$. See Figure 4.4. In this case

$$f^+(\lambda) = \lambda \qquad \text{if } \lambda \geqslant \bar{\lambda}^+$$
$$= \lambda_0^+ + \lambda(\bar{\lambda}^+ - \lambda_0^+)/\bar{\lambda}^+ \qquad \text{otherwise}$$

Case 5: $\underline{a} - \alpha \leqslant \underline{a}' - \alpha'$, $\underline{a} \leqslant \underline{a}'$. In this case $f^+(\lambda) = \lambda$.
The calculation of f^- is similar, changing $\underline{a}$ into $\bar{a}$, $\underline{a}'$ into $\bar{a}'$, α into

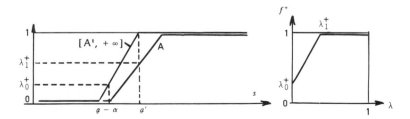

Figure 4.2. Case 2.

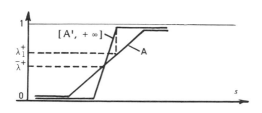
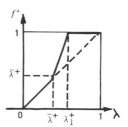

Figure 4.3. Case 3.

β, α' into β', and reversing the inequalities. The quantities λ_0^-, λ_1^-, and $\bar{\lambda}^-$ are evaluated similarly to λ_0^+, λ_1^+, and $\bar{\lambda}^+$.

Hence both f^+ and f^- are piecewise linear; therefore so also is f (though possibly involving more breakpoints).

In order to simplify the calculations, only an approximation to f is computed, $f^\#$ say, such that $f^\# \geqslant f$, and $f^\#$ is of a standard shape. $f^\#$ is defined as follows.

Note first that

$$
\begin{array}{llll}
\text{in case 1,} & \lambda_0^+ = 1, & \lambda_1^+ = 0, & \bar{\lambda}^+ \notin \;]0, 1[\\
\text{in case 2,} & \bar{\lambda}^+ \notin \;]0, 1[& & \\
\text{in case 3,} & \lambda_0^+ = 0 & & \\
\text{in case 4,} & \lambda_1^+ = 1 & & \\
\text{in case 5,} & \lambda_0^+ = 0, & \lambda_1^+ = 1, & \bar{\lambda}^+ \notin \;]0, 1[
\end{array}
$$

so that the quantities λ_0^ε, λ_1^ε, $\bar{\lambda}^\varepsilon$, $\varepsilon \in \{-, +\}$ are always defined. Let $\lambda_0 = \max(\lambda_0^+, \lambda_0^-)$; $\lambda_1 = \min(\lambda_1^+, \lambda_1^-)$. The following tests are made to specify the shape of $f^\#$:

1. If $\lambda_0 = 1$ or $\lambda_1 = 0$ then $f^\# = f$ and $f^\#(\lambda) = 1$, as in case 1. This case results in total uncertainty about the conclusion.

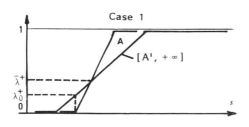
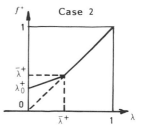

Figure 4.4. Case 4.

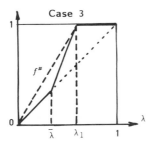 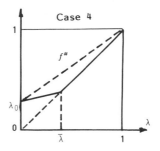

Figure 4.5. Approximating f.

2. If $0 < \lambda_0 < 1$ and $0 < \lambda_1 < 1$ then $f^{\#} \geqslant f$ is defined as f^+ in case 2 (omitting the $+$).

3. If $\lambda_0 = 0$ and $0 < \lambda_1 < 1$, then set $\bar{\lambda} = \min(\bar{\lambda}^+, \bar{\lambda}^-)$; $f^{\#}$ can be defined as f^+ in case 3 (omitting the $+$). We obviate the need to compute $\bar{\lambda}$ by setting it equal to 0 (i.e., the dotted line in Figure 4.5 above).

4. If $\lambda_1 = 1$ and $0 < \lambda_0 < 1$, then set $\bar{\lambda} = \max(\bar{\lambda}^+, \bar{\lambda}^-)$; $f^{\#}$ can be defined as f^+ in case 4 (omitting the $+$). We obviate the need to compute $\bar{\lambda}$ by setting it equal to λ_1 (i.e. the dotted line in Figure 4.5 above).

5. If $\lambda_1 = 1$ and $\lambda_0 = 0$ then $f = f^{\#}$ as in case 5. This case corresponds to $A' \subseteq A$, i.e. $B' = B$.

Finally, $f^{\#}$ is characterized by the pair (λ_0, λ_1), which induces a simple transformation of B into $B^{\#}$ by inference, in any of the cases 2, 3, or 4.

Let $B^{\#}$ be defined by $\mu_{B^{\#}}(t) = f^{\#}(\mu_B(t))$. $B^{\#}$ is of a trapezoidal shape, with a level of uncertainty Θ, as pictured below (Figure 4.6). One advantage of the proposed procedure is that it always produces results of this form. $B^{\#}$ is symbolized by the 5-tuple $(\underline{b}', \bar{b}, \underline{c}', \bar{c}', \Theta)$. It relates to

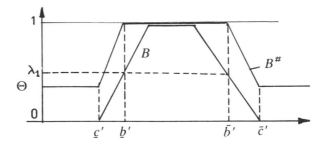

Figure 4.6. Calculation of $B^{\#}$.

the exact result B' as follows:

- $B^\# \supseteq B'$ since $f^\# \geqslant f$, i.e., deducing $B^\#$ instead of B' is logically valid.
- $\dot{B}^\# = \dot{B}'$; indeed the core of B' is determined by the set $\{\lambda \mid f(\lambda) = 1\}$ whose lower bound is clearly $\min(\lambda_1^+, \lambda_1^-)$, i.e., the same lower bound as $\{\lambda \mid f^\#(\lambda) = 1\}$ by construction.
- $\inf \mu_{B'} = \inf \mu_{B^\#} = \lambda_0$. Indeed,

$$\inf\{f(\lambda) \mid \lambda \in [0, 1]\} = \lambda_0 = \inf\{f^\#(\lambda) \mid \lambda \in [0, 1]\} = \Theta$$

is the level of uncertainty of B'.
- $B_\Theta^\# = B_\Theta' =]\underline{c}', \bar{c}'[$, i.e., the points where $\mu_{B^\#}$ attains its level of uncertainty are the same as for B', i.e., they coincide with the end points of the support of B.

Consequently, $\mu_{B^\#}$ differs from $\mu_{B'}$ only on $B_\Theta^\# - \dot{B}^\#$. Let $B = (\underline{b}, \bar{b}, \gamma, \delta)_{LL}$ with $L(x) = \max(0, 1 - x)$. Then the parameters describing $B^\#$ are $\Theta = \lambda_0$, $\underline{c}' = \underline{b} - \gamma$, $\underline{b}' = \underline{b} - \gamma(1 - \lambda_1)$, $\bar{b}' = \bar{b} + \delta(1 - \lambda_1)$, $\bar{c}' = \bar{b} + \delta$.

In the case of noninteractive complex conditions in the rule, i.e., $A = A_1 \times A_2 \times \cdots \times A_n$ and $S = S_1 \times S_2 \times \cdots \times S_n$, let $\Pi_1', \ldots, \Pi_n'$ be the possibility measures associated with $A_1', \ldots, A_n'$, where the fact $A' = A_1' \times A_2' \times \cdots \times A_n'$. It is easily verified that

$$\Pi'(\overline{A_\lambda}) = \Pi'(\overline{A_{1\bar{\lambda}} \times A_{2\bar{\lambda}} \times \cdots \times A_{n\bar{\lambda}}})$$

$$= \max_{i=1,n} \Pi_i'(\overline{A_{i\bar{\lambda}}})$$

Hence, if f_i denotes the modifier of μ_B associated with the rule $A_i \to B$ and the fact A_i', then

$$\forall \lambda, \quad f(\lambda) = \max_{i=1,n} f_i(\lambda)$$

and the mapping $f^\#$ is thus easily obtained from $f_i^\#$, $i = 1, n$. The pair (λ_0, λ_1) associated with $f^\#$ is easily calculated from the $(\lambda_0^i, \lambda_1^i)$ associated with $f_i^\#$ as follows:

$$\lambda_0 = \max_{i=1,n} \lambda_0^i, \quad \lambda_1 = \min_{i=1,n} \lambda_1^i$$

Finally, if the fact A_ω' is of the form $\mu_{A_\omega'} = \max(\mu_{A'}, \omega)$ then the mapping f_ω which modifies B when the rule $A \to B$ is applied is defined by $f_\omega = \max(\omega, f)$. Hence $\mu_{B^\#_\omega} = f^\#_\omega \circ \mu_B$ is defined by the same 5-tuple as $\mu_{B^\#}$ except that $\Theta = \max(\omega, \lambda_0)$, where λ_0 is calculated from A' only. This remark is useful when chaining two rules involving fuzzy predicates.

Appendix B: Computer Programs

```
                            LELISP CODE OF THE SPII SYSTEM

(defun action (acts)
    (cond ((null acts) )
          ((dejadsbf (get (caar acts) 'pbf)) (action (cdr acts)))
          (t (increm) (action (cdr acts)))))
;For executing the 'then' part (represented by "acts") of a rule

(defun dejadsbf (x)
(let ((o (cadar acts))(nf (car x))(f nil)(v nil)(v2 nil)
      (cv nil)(vfonc nil)(args nil)(i nil)(fuzv nil))
(cond
  ((null x) nil)
  ((equal (setq i (infer)) '(0 1)))
  ((eq (cadr (setq f (eval (car x)))) o)
   (if (null (get (car x) 'exhausted))
       (cond((eq '~ (atomcar (caddr f))) (setq v2
               (cond((listp (setq vfonc (caddar acts)))
                      (setq args (mapcar
                         '(lambda (x)(if (eq '? (atomcar x))(car (cassq x jeu)) x))
                                      (cdr vfonc)))
                      (cond ((apply 'or
                              (mapcar '(lambda (x)(eq '~ (atomcar x))) args))
                             (setq v (gen '~))
                             (set v (funcall (get (car vfonc) 'ffloue) args))
                             v)
                            (t (apply (car vfonc) args))))
                     ((cadr (assq vfonc jeu)))
                     (t vfonc)) )
              (setq cv (gen '~))
              (set cv (car (setq fuzv
                 (combf (list (list (caddr f)
                                    (differ 1 (car (get (car x) 'cert))))
                              (list v2 (differ 1 (car i)))))))))
              (set (car x) (list (caar acts) o cv))
              (putprop (car x) (list (differ 1 (cadr fuzv)) 1) 'cert)
              (sauvejeu (car x) nrg)
              (putprop (car x) (append1 (get (car x) 'provient)
                                        (mcons nrg cv i)) 'provient))
             ((if (not (equal (caddar acts)(caddr (eval (car x)))))
               (print "Remise en question de "(car x)" par "nrg)
               (sauvejeu (car x) nrg)
               (putprop (car x) (append1 (get (car x) 'provient)
                                         (cons nrg i)) 'provient)
               (putprop (car x) (comb) 'cert)))))
   (t (dejadsbf (cdr x))))))
;The argument is a list of facts having the same attribute as the current
;conclusion. Returns nil if no fact concerning the same attribute and the same
;object as the current action of "acts" is already in the factual base ("bf").
;In the other case, this function computes (eventually) the value component
;produced by the current rule and performs the combination (i.e. intersection of
;possibility distributions via the "combf" function) of this value with the ones
;previously established. This function is also used to update the properties
;"cert" (certainty associated to a fact) and "provient" (origins of a fact).
;The first updating may call "comb" that combines necessity-possibility pairs.
;The latter property is a list either of the form (rule_of_origin necessity
;possibility) or of the form (rule_of_origin fuzzy_value necessity possibility)
;in case the value component is imprecise.
;NB: the variable "jeu" contains the variable-instantiation pairs.

(de sauvejeu (f nrg) (if jeu (putprop f jeu nrg) nil))
```

```
;Saves the variable-instantiation pairs (i.e."jeu") used when firing the rule
;"nrg" for producing the fact "f".

(defun infer ()
(slet ((z (reverse (car acts)))(s (getsoun (cadr z)))
       (n (getsoun (car z)))(i nil))
   (cond ((or (minusp n) (minusp s))
             (list (differ 1 (cadr (setq i
                       (calinfer (times -1 s)(times -1 n)))))
                   (differ 1 (car i))))
         (t (calinfer s n)))))
;Returns the necessity-possibility pair representing the uncertainty of the
;inferred conclusion. If "s" or "n" is a negative number it means that the
;conclusion to be inferred is the negation of the triple (the first three
;elements of (car acts)) specified in the rule.
;"s" and "n" may be obtained (via getsoun) from a table in case of rules
;of the form "the more X is A the more certain q"

(defun calinfer (s n)
(list  (apply 'minr (list s (car certtp)))
       (differ 1 (apply 'minr (list n (differ 1 (cadr certtp)))))))
;numerical computation of necessity_possibility pairs given "s", "n" and the
;pair "certtp" expressing the satisfying of the condition part.

(defun increm ()
(let ((x nil)(v nil)(i nil)(atto (firstn 2 (car acts)))
      (vfonc nil)(args nil)(vinst nil))
(if (equal '(0 1) (setq i (infer)))
    nil
    (setq x (gen 'f))(putprop x i 'cert)
    (cond ((listp (setq vfonc (caddar acts)))
              (setq args   (mapcar
                  '(lambda (x) (cond ((eq '? (atomcar x))(car (cassq x jeu)))
                                     (t x)))
                  (cdr vfonc)))
              (set x (append1 atto
                  (cond ((apply 'or
                            (mapcar '(lambda (x)(eq '~ (atomcar x))) args))
                         (setq v (gen '~))
                         (set v (funcall (get (car vfonc) 'ffloue) args))
                         v)
                        (t (apply (car vfonc) args))))))
          ((setq vinst (assq vfonc jeu)) (set x (nconc atto (cdr vinst))))
          (t (set x (append1 atto vfonc))))
    (if (eq '~ (atomcar (setq v (caddr (eval x)))))
        (putprop x (list (mcons nrg v i)) 'provient)
        (putprop x (list (cons nrg i)) 'provient))
    (newr bf x)(sauvejeu x nrg)
    (putprop (caar acts) (append (get (caar acts) 'pbf) (list x)) 'pbf))))
;Production of a new fact that is put in the factual base.
;Works similarly to "dejadsbf".
;"i" is the nessecity-possibility pair. The value component may be computed.
;The property "pbf" is an index permitting the grouping of facts having the same
;attribute.

(de gen (x)
 (cond ((eq x 'f)
          (implode (cons (cascii x) (explode (setq comptf (1+ comptf))))))
        ((eq x '~)
          (implode (cons (cascii x) (explode (setq compt~ (1+ compt~))))))
        (t (print "erreur ds 'gen' ")))
;Generates an atom of the type f12 denoting a fact or
;generates an atom of the type ~32 denoting a fuzzy value

(defun demontr (triple)
(slet ((flag nil) (non nil)(relevant nil)(v nil)(cert nil)
       (vt (caddr triple))(atcarv (atomcar vt))(e-y-n-exhaust nil))
   (cond ((eq 'non (car triple))
            (setq non (let ((triple (cadr triple))) (demontr triple)))
            (list (differ 1 (cadr non))(differ 1 (car non))))
         ((exist (get (car triple) 'pbf)))
         ((null
            (if (null e-y-n-exhaust)
                (setq relevant (get (car triple) 'crosref))
                (setq relevant (get (car triple) 'crosref))
                (do ((y (mapcar '(lambda (x)(car x))
                            (get e-y-n-exhaust 'provient)) (cdr y)))
```

```
                              ((null y) relevant)
                              (setq relevant (remq (car y) relevant)))))
               '(0 1))
              ((eq '? atcarv)
                (cond ((setq cert (ttrg (rgnonfloue relevant)))
                       (putprop nf t 'exhausted)
                       (newr jeu (list vt (setq cv (caddr (eval nf)))))
                       cert)
                      ((setq v (demontrf triple (rgfloue relevant)) cv v)
                       (putprop nf t 'exhausted) (newr jeu (list vt v))
                       (if (eq cv '~indetermine)
                           '(0 1)
                           (get nf 'cert)))))
              ((eq '~ atcarv)
                  (cond ((setq cert (ttrg (rgnonfloue relevant)))
                         (putprop nf t 'exhausted)
                         (cond ((numberp (setq cv (caddr (eval nf))))
                                (setq pos (mu cv (eval vt))
                                      nec  pos))
                               (t (setq nec (neccc (eval vt)(eval  cv))
                                        pos (poscc (eval vt)(eval  cv)))))
                         (cond ((not (equal cert '(1 1)))
                                (normalise
                                    (if (> nec (car cert)) (car cert) nec)
                                    (if (> (- 1 pos) (car cert))
                                        (- 1 (car cert)) pos)))
                               ((normalise nec pos))))
                        ((setq v (demontrf triple (rgfloue relevant)) cv v)
                         (putprop nf t 'exhausted)
                         (if (eq (eval v) '~indetermine) '(0 1)
                             (setq nec (neccc (eval  vt) (eval v))
                                   pos (poscc (eval  vt) (eval v))
                                   cert (get nf 'cert))
                             (normalise
                                 (if (> nec (car cert)) (car cert) nec)
                                 (if (> (- 1 pos) (car cert))
                                     (- 1 (car cert)) pos))))
                        (t '(0 1))))
              ((if (and (ttrg (rgnonfloue relevant))(putprop nf t 'exhausted))
                   (if (or (if (listp vt) (member (caddr (eval nf)) vt) nil)
                           (equal vt (caddr (eval nf))))
                       (get nf 'cert)
                       nil)
                   nil))
              (t '(0 1)))))
;Return the necessity_possibility pair representing the satisfying
;of "triple". In case this satifying involves a fuzzy pattern matching
;a normalization may occur. If no fact (in the fact base) satisfies the goal
;"triple", then one tries to establish it by backward chaining, that is by
;using rules of "relevant" (i.e.rules concluding on the attribute of "triple").
;Rules with sufficiency and necessity degrees are used first (the others,
;that are applied with the generalized modus ponens technique, are
;used only if none of the first kind has been fired).
;vt=value component of "triple", "cv" or "v" is the value component produced
;by firing a rule ("v" is local to the function, whereas "cv" is also used in
;the "ttpremf" function). In case the argument is the negation of an elementary
;information item (i.e. a triple) the returned nec-pos pair is the negation
;of the satisfying of the triple

(de apropos (triple)
(let ((jeu nil) (nf nil) (atcarv (atomcar (caddr triple)))
      (flag nil) (cert nil) (rgnf nil))
  (cond ((not (eq atcarv '?))(print "certitude =  " (demontr triple)))
        ((setq cert (exist (get (car triple) 'pbf)))
         (print "certitude =  " cert)
         (print (caar jeu) " = " (cadar jeu))(terpri))
        ((null (setq relevant (get (car triple) 'crosref)))
         (print triple " n'est pas demontrable avec les regles disponibles"))
        ((and (setq rgnf (rgnonfloue relevant))(setq cert (ttrg rgnf)))
         (putprop nf t 'exhausted) (print "certitude =  " cert)
         (print (caddr triple) " = "
                (caddr (eval (findnf (get (car triple) 'pbf))))(terpri))
        ((setq v (demontrf triple (rgfloue relevant)))
         (putprop nf t 'exhausted) (print "certitude =  " (get nf 'cert))
         (print (caddr triple) " = " v ) (terpri))
        ((print triple " n'est pas demontrable")))))
```

```
;top-level query function. Returns the certainty and eventually the fuzzy value
;component of the triple being queried about.

(de rgfloue (rel) (mapcan '(lambda(r) (if (get r 'floue) (list r) nil)) rel))
(de rgnonfloue(rel) (mapcan '(lambda(r) (if (get r 'floue)  nil (list r))) rel))
;"rgnonfloue" selects rules of "rel" involving sufficiency and necessity degrees
;"rgfloue" selects the others (to be used with the generalized modus ponens).

(defun exist (faits)
(slet ((x nil) (vt (caddr triple))(atcarv (atomcar vt))
      (cert nil)(nec nil)(pos nil)(f nil)(vinst nil))
  (cond ((and (not (eq '~ atcarv)) (null faits)) nil)
       ((eq '~ atcarv)
           (cond ((eq (cadr triple)(cadr (setq  f (eval (car faits)))))
                 (if (null (get (car faits) 'exhausted))
                   (setq e-y-n-exhaust (car faits) cert nil)
                   (cond ((numberp (setq cv (caddr f)))
                          (setq pos (mu cv (eval vt)) nec  pos))
                         (t (setq nec (neccc (eval vt)(eval cv))
                            pos (poscc (eval vt)(eval cv)))))
                   (cond ((not(or(null (setq cert (get (car faits) 'cert)))
                                 (equal cert  '(1 1))))
                           (normalise
                              (if () nec (car cert)) (car cert) nec)
                              (if () (- 1 pos) (car cert))
                                 (- 1 (car cert)) pos)))
                         ((normalise nec pos)))))
                 ((cdr faits)(exist (cdr faits)))))
       ((eq (cadr (setq f (eval (car faits)))) (cadr triple))
         (if (null (get (car faits) 'exhausted))
           (setq e-y-n-exhaust (car faits) cert nil)
           (cond ((listp vt)
                  (cond ((memq (setq cv (caddr f)) vt)
                         (cond ((get (car faits) 'cert))
                               (t '(1 1))))))
                 ((eq atcarv '?)
                  (cond ((setq vinst (assq vt jeu))
                         (if (eq (cadr vinst)(setq cv (caddr f)))
                             (cond ((get (car faits) 'cert))
                                   (t '(1 1)))
                           nil))
                        ((newr jeu (list vt (setq cv (caddr f))))
                         (cond ((get (car faits) 'cert))
                               (t '(1 1))))))
                 ((eq vt (setq cv (caddr f)))
                  (cond ((get (car faits) 'cert))
                        (t '(1 1))))))
         (t (exist (cdr faits)))))))
;Returns the necessity-possibility pair representing the satisfying of the
;goal "triple". Returns nil if no fact (of "facts") matches "triple".
;A fact is "exhausted" when it is an initial one or when application of
;all rules concluding on its attribute component has been attempted.
;If "triple" is imprecise (i.e. its value component starts with ~) the nec-pos
;pair is obtained via fuzzy pattern matching (which involves a normalization
;process). If "triple" has a value component constituted by a list, then the
;value component of the fact matched against it has to be one of the above list.
;By default, imprecise facts given initially by the user are assumed to be
;certain. "cv" is the value component of the fact matched against "triple"
;"cv" is used in ttpremf.

(defun normalise (nec pos)
(let ((d (maxr (difference 1 nec) pos)))
  (cond ((equal d pos) (list (divide (difference (plus d nec) 1) d) 1))
        (t (list 0 (divide pos d))))))
;Normalizes a (nec pos) pair, where "nec" and "pos" are both in ]0,1[, into
;a pair of the type [0,npos] or [nnec,1] reflecting whether the information
;being conveyed is rather false or rather true respectively.

(defun findnf (l)
   (cond ((null l) (print "findnf ne trouve pas" triple) nil)
         ((equal (firstn 2 (eval (car l)))(firstn 2  triple))
          (car l))
         (t (findnf (cdr l)))))
;For retrieving the name (i.e.f34) of a fact given as a triple

(defun ttrg (relev)
```

```
(cond ((null relev)
            (if flag (get (setq nf (findnf (get (car triple) 'pbf))) 'cert) nil))
         ((member (firstn 2 triple) (mapcar '(lambda (x) (firstn 2 x))
                           (cadr (eval (car relev)))))
           (cond ((essayer (car relev)) (setq flag t) (ttrg (cdr relev)))
                      (t (ttrg (cdr relev)))))
         (t (ttrg (cdr relev)))))
;Tries to apply all rules of "relev" that conclude on the first two
;components of "triple".
;In case one of these rules has been fired, it returns the nec-pos pair of
;the related deduced fact. Returns nil otherwise. "flag" is put to t as soon
;as one rule has been fired.

(defun essayer (nrg)
(let ((x (eval nrg))(certtp nil)(jeu nil))
    (cond ((equal (setq certtp (ttprem (car x))) '(0 1)) nil)
             (t (action (cadr x))))))
;The rule denoted by "nrg" is fired if the condition part is satisfied
;with a nec-pos pair different from (0 1).

(defun ttprem (ensprem)
(slet ((cv nil) (i (demontr (car ensprem))) (i1 nil))
    (cond ((null (cdr ensprem)) i)
             ((equal '(0 0) i) '(0 0))
             (t (list (minr (car i)(car (setq i1 (ttprem (cdr ensprem)))))
                        (minr (cadr i)(cadr i1)))))))
;Returns the nec-pos pair representing the satisfying of a conjunction
;of premises. In case information about a premise is not readily available
;"demontr" will ensure appropriate invocation of rules (otherwise "demontr"
;returns directly the desired nec-pos pair).

(defun demontrf (triple relevant)
(let ((f nil)(lval nil)(prov nil)(v nil)(z nil)(jeu1 nil))
    (cond ((null relevant) nil)
             ((setq z (combf (mapcar
                           '(lambda (x) (if (eq (car x) '~indetermine) nil (list x)))
                           (setq lval (ttrgf relevant)))))
               (setq f (gen 'f) nf f v (gen '~)) (set v (car z))
               (set f (list (car triple)(cadr triple) v)) (newr bf f)
               (let ((jeu jeu1)(mapcar '(lambda (rgx) (sauvejeu f rgx)) relevant))
               (putprop (car triple) (append1 (get (car triple) 'pbf) f) 'pbf)
               (putprop f (list (differ 1 (cadr z))1)  'cert)
               (putprop f
                          ((lambda (l1 l2)
                              (while l1 (newr prov (mcons (car l1)(caar l2)
                                                             (list (differ 1 (cadar l2)) 1))
                                             (nextl l1)(nextl l2)) prov) relevant lval)
                          'provient)
                    v))))
;Returns the imprecise value component (or more specifically the atom denoting
;the possibility distribution representing this imprecise value) to be matched
;against the component value of "triple".
;It first looks for this value in existing facts. If no fact having the
;attribute and the object of "triple" exists, it tries to get it by backward
;chaining (using "ttrgf") on rules to be applied with the generalized modus
;ponens. When several rules are fired, a combination (using "combf") must take
;place. Updates properties linking attributes to facts having them (i.e. "pbf")
;and facts to their origins (i.e."provient"). The latter property is a list of
;the form ((r1 ~v1 N1 Π)...(rn ~vn Nn Πn)) where the ri's are the rules that
;have been used, ~vi are imprecise values, and the Ni Πi are necessity and
;possibility degrees.

(defun ttrgf (relev)
    (cond ((null relev) nil)
             ((equal (firstn 2 triple) (firstn 2 (caadr (eval (car relev)))))
               (cons (apprgf (car relev))(ttrgf (cdr relev))))
             (t (ttrgf (cdr relev)))))
;Tries to apply all rules of "relev" that conclude on the first two
;components of "triple" by using the generalized modus ponens technique.
;Returns the list of pairs of the form val-imposs where val is a distribution
;imposs is the degree of impossibility (i.e. 1-necessity) attached to this
;distribution. Each pair corresponds to one rule of "relev".

(defun ttpremf (ensprem)
(let ((cv nil) (v~ nil) (cert nil) (f nil) (triple (car ensprem)))
```

```
(cond ((or (null ensprem)(null zf)) nil)
      ((if (setq zf (and (listp (caddr conc))
                         (memq (caddr triple)(caddr conc))))
           t
           (newr aux (caddr triple)))
       (cond (zf (demontr triple)
              (if (or (null (setq f (findnf (get (car triple) 'pbf))))
                      (equal '(0 1) (get f 'cert)))
                  nil
                  (ttpremf (cdr ensprem))))
             (t (setq zf (demontr triple))
              (if (or (null (setq f (findnf (get (car triple) 'pbf))))
                      (equal '(0 1) (setq cert (get f 'cert)))
                      (< (cadr cert) 1)
                      (and (not (eq '~ (atomcar (caddr triple))))
                           (not (equal (caddr triple) cv))))
                  nil
                  (cons (cond ((null cert) (list cv 0))
                              (t (list cv (differ 1 (car cert)))))
                        (ttpremf (cdr ensprem))))))))))
;Used when applying the generalized modus ponens technique.
;Puts in "aux" the value components of conditions of the rule considered
;(puts only those that are not appearing has parameter in the conclusion)
;When information about a premise is not available, "ttpremf" makes a call
;to "demontr" (itself calling "demontrf" if needed).
;Returns the list (corresponding one to one with "aux") of pairs of the form
;val-imposs where val is the trapezoid which is the value component of the
;fact matching the premise associated to the corresponding element in "aux",
;and imposs is the degree of impossibility of this fact (i.e. 1-necessity).
;If zf becomes nil, it is impossible to match the current premise (it is then
;useless to go further with the rule under consideration).

(defun apprgf (nrg)
  (slet ((x (eval nrg)) (jeu nil) (zf t) (aux nil) (lambda0 nil)
         (conc (caadr x)) (z nil) (v nil)
         (a0 (ttpremf (car x))) (a aux)
         (ce (eval (caddr conc))))
  (setq z (mpg))
  (if (eq '~indetermine z)
      '(~indetermine 0)
      (setq v (gen '~) jeu1 jeu) (set v z) (list v lambda0)))))
;Deduction of a possibility distribution (i.e.trapezoid together with a  degree
;of impossibility (i.e.1-necessity)) characterizing the inferred item by calling
;the fast generalized modus ponens (GMP). Returns the list constituted of this
;trapezoid and this impossibility degree.
;"a" = list of distribution (trapezoid) involved in premises,
;"a0" = list of pairs (trapezoid impossibility_degree) involved in facts
;(one-to-one correspondence with "a"), "nrg" = rule name, "x" = the rule itself.
(defun mu (x quadr)
  (cond ((and (<= x (cadr quadr))(<= (car quadr) x)) 1)
        ((or (<= (plus (cadddr quadr) (cadr quadr)) x)
             (<= x (differ (car quadr) (cadddr quadr)))) 0)
        ((< x (car quadr))
         (divide (plus x (caddr quadr) (* -1 (car quadr))) (cadddr quadr)))
        (t (divide (plus (* -1 x) (cadr quadr) (cadddr quadr))
                   (cadddr quadr)))))
;membership degree of "x" in the fuzzy set (or possibility distribution)
;defined by the 4-tuple "quadr".


(defun mpg ()
  (let ((lambda1 nil))
  (cond ((= 1 (setq lambda0 (apply 'maxr (mapcar 'delta1 a a0))))
         '~indetermine)
        ((= 1 (setq lambda1 (apply 'minr (mapcar 'locallambda1 a a0)))) ce)
        ((list (differ (car ce) (times (caddr ce)(differ 1 lambda1)))
               (plus (cadr ce) (times (cadddr ce)(differ 1 lambda1)))
               (divide (times lambda1 (caddr ce))(differ 1 lambda0))
               (divide (times lambda1 (cadddr ce))(differ 1 lambda0)))))))
;Fast generalized  modus ponens (FGMP).
;"mpg" computes an approximation of what should really be inferred by a strict
;application of the generalized modus ponens deduction technique. This
;approximation is described in Appendix A of this chapter.
;Returns a four-element list representing the deduced distribution and computes
;"lambda0" which is the global level of indetermination attached to it.
;"lambda1" is the degree of inclusion of the core of the compound distribution
```

```
;composed from those involved in the facts, with the parallel compound
;distribution constructed from the premises. (See Appendix A for more on
;"lambda0" and "lambda1" which play a key role in computing the FGMP)
;"~indetermine" denotes a distribution conveying complete indetermination.
;Recall that in this version of SPII a collection of fuzzy rules (describing the
;dependency between given variables) are not combined before the inference but
;rather each rule is fired and then a combination of the different results is
;performed.

(defun delta1(a1 a01)
(let ((u1 nil)(u2 nil)(a1e nil)(a01e nil))
(cond ((null a01) (print "err ds delta1"))
      ((not (eq '~ (atomcar (car a01)))) (cadr a01))
      ((not (eq '~ (atomcar a1))) 1)
      ((setq u1 (differ (car (setq a1e (eval a1)))(caddr a1e))
            u2 (plus (cadr a1e)(cadddr a1e)))
          (cond ((or (< (car (setq a01e (eval (car a01)))) u1)
                     (< u2 (cadr a01e))) 1)
                ((zerop (cadddr a01e))
                   (cond ((zerop (caddr a01e)) (cadr a01))
                         ((maxr (mu u1 a01e)(cadr a01)))))
                ((zerop (caddr a01e))(maxr (mu u2 a01e)(cadr a01)))
                ((maxr (mu u1 a01e)(mu u2 a01e)(cadr a01)))))))
;Computation of the global level of indetermination arising from the
;confrontation of "a1" (a distribution in a premise) against "a01" (the
;distribution and its associated impossibility degree in the corresponding
;fact).

(defun locallambda1 (a1 a01)
(let ((a1e nil)(a01e nil))
(cond ((null a01) (print "err ds locallambda1"))
      ((not (eq '~ (atomcar (car a01)))) 1)
      ((setq a01e (eval (car a01)))(setq a1e (eval a1))
       (minr (mu (car a01e) a1e)(mu (cadr a01e) a1e))))))
;Returns the degree of inclusion of the core of distribution associated to a
;paticular fact in the distribution associated to the corresponding premise.


(de combf (liste)
(let ((x nil)(~ nil))
 (if liste
     (cons (eval (car (setq x (combf1 (sort '(lambda(a b)(<= (cadr a)(cadr b)))
                                              liste)))))
           (cdr x)))))

(defun combf1 (liste)
   (cond ((null (cdr liste)) (car liste))
         (t (inter (car liste) (combf1 (cdr liste))))))
;Combination of pairs (trapezoidal_distribution , impossibility) of "liste"

(defun inter (p1 p2)
(let ((p1e nil)(p2e nil)(u1l nil)(u1r nil)(u2l nil)(u2r nil)(xi nil))
(cond
 ((eq (car p2) '~indetermine) p1)
 ((setq p1e (eval (car p1)) p2e (eval (car p2)))
  (setq u1r (plus (cadr p1e)(cadddr p1e)) u2l (differ (car p2e)(caddr p2e)))
  (setq u2r (plus (cadr p2e)(cadddr p2e)) u1l (differ (car p1e)(caddr p1e)))
  (cond((and (= (cadr p1)(cadr p2)) ;same degrees of indetermination
             (or (>= (car p2e)(cadr p1e))(>= (car p1e)(cadr p2e)));disjointcores
             (or (<= u1r u2l)(<= u2r u1l) ;disjoint supports
                 (and (< u2l u1r) ;non-disjoint support but the greater degree
                          ;of intersection of p1e with p2e is < indetermination
                      (setq xi (divide (plus (times u2l (cadddr p1e))
                                             (times u1r (cadddr p2e)))
                                       (plus (cadddr p2e)(cadddr p1e))))
                      (< (mu xi p1e) (cadr p1)))
                 (and (< u1l u2r) ;non-disjoint support but the greater degree
                          ;of intersection of p1e with p2e is < indetermination
                      (setq xi (divide (plus (times u1l (cadddr p2e))
                                             (times u2r (cadddr p1e)))
                                       (plus (cadddr p1e)(cadddr p2e))))
                      (< (mu xi p1e) (cadr p1)))))
        (if (equal p1e p2e) p1 '('~indetermine 0)))
       ((>= (car p2e)(cadr p1e)) ;disjoint cores
        (empty-inter-cores p1 p2))
       ((>= (car p1e)(cadr p2e)) ;disjoint cores
```

```
                (empty-inter-cores p2 p1))
            (((< (car p1e)(car p2e))                    ;non-disjoint cores
                (noempty-inter-cores (cons p1e (cdr p1))(cons p2e (cdr p2))))
            ((noempty-inter-cores (cons p2e (cdr p2))(cons p1e (cdr p1)))))))))
                ;non-disjoint cores
;Performs the combination of the information items p1 , p2. This
;combination is an approximation of a renormalized fuzzy set intersection.
;The renormalization takes place when the distributions involved in p1 and
;p2 have disjoint cores. An approximation occurs when the real intersection
;does return an information item having the form of a trapezoidal distribution
;qualified by a degree of indetermination. The implemented approximation
;process returns a pair of the form (trapezoid indetermination) in which the
;trapezoid is (not exactly but very closely) the smaller trapezoid containing
;the distribution that would be obtained by combining the trapezoid
;involved in p1 and p2 without approximation. Therefore, this approximation
;returns a valid but less specific result than the plain intersection.
;p1 and p2 are pairs each composed of a trapezoidal distribution and
;its associated level of indetermination (or impossibility degree).
;Returns a list composed of a 4-tuple representing a trapezoidal
;distribution and a scalar value conveying the impossibility
;degree attached to this distribution.

(de empty-inter-cores (p1 p2)
(slet ((p1e (eval (car p1)))(p2e (eval (car p2)))
       (yi nil)(xi nil)(core nil)(delta nil)
       (u1r (plus (cadr p1e)(cadddr p1e)))(u2l (differ (car p2e)(caddr p2e)))
       (u2r (plus (cadr p2e)(cadddr p2e)))(u1l (differ (car p1e)(caddr p1e))))
(cond((and (< u2l u1r)
            (setq xi (divide (plus (times u2l (cadddr p1e))
                                   (times u1r (cadddr p2e)))
                             (plus (cadddr p2e)(cadddr p1e))))
            (> (mu xi p2e) (maxr (cadr p1)(cadr p2))))   ;the greater degree of
                ;intersection of p1e with p2e is > than the greater
                ;degree of indetermination
         (cond ((<= (cadr p1)(cadr p2))
                (setq yi (mu xi p1e) core (list xi xi))
                (setq ~
                  (nconc core
                    (list
                     (if (= (cadr p1)(cadr p2))
                         (divide (differ xi (car (mu-1 (cadr p1) p2e)))
                                 (differ 1 (setq delta (divide (cadr p1) yi))))
                         (divide (differ xi (car (mu-1 (cadr p1) p1e)))
                                 (differ 1 (setq delta (divide (cadr p1) yi)))))
                     (divide (differ (cadr (mu-1 (cadr p1) p1e)) xi)
                             (differ 1 delta)))))
                (list '~ delta))
              (t (setq yi (mu xi p1e) core (list xi xi))
                (setq ~
                  (nconc core
                    (list
                     (divide (differ xi (car (mu-1 (cadr p2) p2e)))
                             (differ 1 (setq delta (divide (cadr p2) yi))))
                     (divide (differ (cadr (mu-1 (cadr p2) p2e)) xi)
                             (differ 1 delta)))))
                (list '~ delta))))
       (t (cond((<= (cadr p1)(cadr p2)) (setq core (mu-1 (cadr p2) p1e))
                (setq ~
                  (nconc core
                    (list
                     (divide (differ (car core)(car (mu-1 (cadr p1) p1e)))
                             (differ 1 (setq delta (divide (cadr p1)(cadr p2)))))
                     (divide (differ (cadr (mu-1 (cadr p1) p1e)) (cadr core))
                             (differ 1 delta)))))
                (list '~ delta))
              (t (setq core (mu-1 (cadr p1) p2e))
                (setq ~
                  (nconc core
                    (list
                     (divide (differ (car core)(car (mu-1 (cadr p2) p2e)))
                             (differ 1 (setq delta (divide (cadr p2)(cadr p1)))))
                     (divide (differ (cadr (mu-1 (cadr p2) p2e)) (cadr core))
                             (differ 1 delta)))))
                (list '~ delta))))))
;Intersection of two distributions (each described by a trapezoid and a degree
;of indetermination (or impossibility degree) having disjoint cores.
```

```
;This function performs an approximation of a renormalized intersection.
;It is assumed that the upper bound of the support of the first trapezoid is
;greater than the lower bound of the support of the second one.
;Returns a list composed of a 4-tuple (trapezoid) and a degree of
;indetermination.

(de noempty-inter-cores (q1 q2)
(let ((p1e (car q1)) (p2e (car q2)))
  (cond (((< (cadr q1)(cadr q2))   ;the second item is more indeterminate
          (setq ~
                (list (car p2e) (minr (cadr p1e)(cadr p2e))
                      (divide (differ (car p2e)(car (mu-1 (cadr q1) p1e)))
                              (differ 1 (cadr q1)))
                      (if (> (cadr p1e) (cadr p2e))
                          (divide (differ (cadr (mu-1 (cadr q1) p1e))
                                          (cadr p2e))
                                  (differ 1 (cadr q1)))
                          (cadddr p1e))))
                (list '~ (cadr q1)))
        ((setq ~ (list (car p2e) (minr (cadr p1e)(cadr p2e))
                       (caddr p2e)
                       (if (> (cadr p1e) (cadr p2e))
                           (cadddr p2e)
                           (if (= (cadr q1)(cadr q2)) (cadddr p1e)
                               (divide (differ (cadr (mu-1 (cadr q2) p2e))
                                               (cadr p1e))
                                       (differ 1 (cadr q2)))))))
                (list '~ (cadr q2)))))))
;Called when the combination to be performed concerns distributions (i.e.
;(car p1) and (car p2)) such that the intersection of their cores is nonempty.
;It is assumed that the lower bound of the core of the second distribution
;belongs to the core of the first one. This function returns a list composed of
;a 4-tuple (trapezoid) and a degree of indetermination.

(defun mu-1 (y quad)
  (cond ((= y 1) (list (car quad)(cadr quad)))
        ((zerop y)(list (differ (car quad)(caddr quad))
                        (plus  (cadr quad)(cadddr quad))))
        ((zerop (cadddr quad))
         (cond ((zerop (caddr quad))(list (car quad)(cadr quad)))
               (t (list (car quad)(plus (times (cadddr quad)
                                               (differ 1 y)(cadr quad)))))))
        ((zerop (cadddr quad))(list (plus (times (caddr quad)
                                                 (differ y 1))(car quad))(cadr quad)))
        (t (list (plus (times (cadddr quad)(differ y 1))(car quad))
                 (plus (times (cadddr quad) (differ 1 y))(cadr quad))))))
;To get the inverse of the degree "y" from the parametrized membership
;function defined as the 4-tuple "quad". Returns a pair of values
;representing either the 2 values having membership equal to "y" or
;the boundaries of an interval of values having the membership degree "y".


(defun comb()
  (let ((negposit (trinegposit (mapcar 'cdr (get nf 'provient)) nil nil)))
    (cond ((null (car negposit))(list (apply 'maxr (cadr negposit)) 1))
          ((null (cadr negposit))(list 0 (apply 'minr (car negposit))))
          (t (normalise (apply 'maxr (cadr negposit))
                        (apply 'minr (car negposit)))))))
;Combines (without reinforcement) nec-pos pairs that are attached to the fact
;"nf" via the property "provient". First, positive and negative contributions
;are combined separately. The results are then summarized via the function
;"normalise".

(defun trinegposit (li neg posit)
  (cond ((null li)(list neg posit))
        ((< 0 (caar li))(trinegposit (cdr li) neg (cons (caar li) posit)))
        (t (trinegposit (cdr li) (cons (cadar li) neg) posit))))
;Separates negative (i.e. of the form (0 pos)) and positive (i.e. of the form
;(nec 1)) nec-pos pairs

(defprop mult *f ffloue)(defprop plus +f ffloue)(defprop differ -f ffloue)
(synonymq mult times)

(de +f (args)
  (cond ((null (cdr args))
```

```lisp
                (if (eq '~ (atomcar (car args))) (eval (car args))
                    (list (car args) (car args) 0 0)))
           (t  (+f2arg
                (if (eq '~ (atomcar (car args))) (eval (car args))
                    (list (car args) (car args) 0 0))
                (+f (cdr args)))))))
;To perform addition of fuzzy or nonfuzzy arguments

(de -f (args)
  (cond ((null (cdr args))
                (if (eq '~ (atomcar (car args))) (eval (car args))
                    (list (car args) (car args) 0 0)))
           (t  (-f2arg
                (if (eq '~ (atomcar (car args))) (eval (car args))
                    (list (car args) (car args) 0 0))
                (-f (cdr args)))))))
;To perform substraction with fuzzy or nonfuzzy arguments

(de *f (args)
  (cond ((null (cdr args))
                (if (eq '~ (atomcar (car args))) (eval (car args))
                    (list (car args) (car args) 0 0)))
           (t  (*f2arg
                (if (eq '~ (atomcar (car args))) (eval (car args))
                    (list (car args) (car args) 0 0))
                (*f (cdr args)))))))
;To perform multiplication of fuzzy or nonfuzzy arguments

(de +f2arg (a1 a2)
  (list (plus (car a1)(car a2)) (plus (cadr a1)(cadr a2))
        (plus (caddr a1)(caddr a2)) (plus (cadddr a1)(cadddr a2))))
;To add the 2 distributions "a1" and  "a2" that are given as trapezoidal ones

(de -f2arg (a1 a2)
  (list (differ (car a1)(cadr a2)) (differ (cadr a1)(car a2))
        (plus (caddr a1)(caddr a2)) (plus (cadddr a1)(cadddr a2))))
;To compute (- a1 a2) with "a1" "a2" given as trapezoidal distributions

(de *f2arg (a1 a2)
  (list (times (car a1)(car a2)) (times (cadr a1)(cadr a2))
        (plus (times (differ (car a1)(caddr a1))(caddr a2))
           (times (car a2)(caddr a1)))
        (plus (times (cadr a2)(cadddr a1))
           (times (plus (cadr a1)(cadddr a1))(cadddr a2)))))
;To multiply the 2 distributions "a1" and "a2" that are given as
;trapezoidal ones



(de getsoun (l)
(cond ((numberp l) l)
      ((eq '- (car l))(* -1 (calsoun (cdr l))))
      ((calsoun l))))
;Returns the appropriate value for "s" or "n"."l" is the last or the last but
;one component of a 5-tuple specifying a conclusion. If this component is a
;number then it is the value we are looking for, otherwise this component is
;a list expressing that "s" or "n" are obtainable by looking in a tableau.
;Recall that the negative sign permits expressing a negative conclusion.

(de calsoun (l)
(let ((intd (get nrg (car l))) (ret nil))
(apply 'minr (intervok
                (mapcar '(lambda (x)
                           (let ((y nil))
                           (cond ((numberp  (setq y (car (cassq x jeu)))))
                                 ((firstn 2 (eval y))))))
                       (cdr l))
                intd))))
;Retrieves the degree of sufficiency or necessity. "intd" is the tableau
;defining what degree (for s ou n) is asssociated to what values (or set of
;values) of the parameters involved. These parameters are variables
;the instantiation of which has to be retrieved from "jeu". Because such an
;instantiation does not necessarily represent a precise numerical value,
;the computation of the degree (for s or n) has to be performed by "intervok".


(de intervok (pnorf id)
```

```
(let ((d nil))
(cond ((null id) (if ret ret '(0)))
      ((setq d (checkint pnorf (car id)))(newl ret d)(intervok pnorf (cdr id)))
      ((intervok pnorf (cdr id))))))
;"id" is the tableau defining what degree (for s or n) is asssociated to
;what values (or set of values) of the parameters involved. "id" (given
;by the expert) is of the form (((a1 b1)(a2 b2) .4) ((a1 b1)(c2 d2) .6)...)
;where the pairs (ai bi) represent intervals (here s or n depends on two
;parameters). "pnorf" is the list of intervals or numbers representing cores
;of the instantiated parameters (which are component values of facts).
;"intervok" returns the list of degrees (for s or n) associated to the elements
;of "id" that have a nonempty intersection with those of "pnorf".
;"calsoun" keeps then the smaller of these degrees (this corresponds
;to a least commitment attitude).

(de checkint (pnorf ints-d)
(cond ((numberp (car ints-d))(car ints-d))
      ((nonvide (car pnorf)(car ints-d))(checkint (cdr pnorf)(cdr ints-d)))))
;"int-d" is an element of "id" (see the function "intervok"), that is, a list
;of intervals plus a degree as the last element.
;"checkint" returns this degree if each of the intervals has a nonempty
;intersection with their corresponding interval in "pnorf".

(de nonvide (pnor int)
(cond ((numberp pnor) (and (>= pnor (car int))(< pnor (cadr int))))
      ((or  (and (>= (car pnor) (car int))(< (car pnor) (cadr int)))
            (and (>= (cadr pnor) (car int))(< (cadr pnor) (cadr int)))))))
;Returns nil in case "pnor" has an empty intersection with "int".


(de atomcar (x) (car (explodech x)))

(de atomcdr (x) (implode (cdr (explode x))))

(dm minr (minr . x)
     (list 'cond
          (list (null  x) (list 'print "err minr"))
          (list (null (cdr x)) (car x))
          (list (list '< (cons 'minr (cdr x)) (car x)))
          (list t (car x))))
;Computes the minimum of the list of values of "x" (either integer or real
;numbers)

(dm maxr (maxr . x)
     (list 'cond
          (list (null  x) (list 'print "err maxr"))
          (list (null (cdr x)) (car x))
          (list (list '> (cons 'maxr (cdr x))(car x)))
          (list t (car x))))
;Computes the maximum of the list of values of "x" (either integer or real
;numbers)


(defun poscc (c1 c2)
(let ((-x (differ (cadr c2) (car c1)))
      (-y (differ (cadr c1) (car c2))))
 (cond ((<= (maxr -x -y)
        (plus (differ (cadr c1)(car c1)) (differ (cadr c2)(car c2)))) 1)
       ((> -x -y)
        (cond ((and (zerop (caddr c2)) (zerop (cadddr c1))) 0)
              (t (born (divide (differ (car c2) (cadr c1))
                               (plus (cadddr c1) (caddr c2)))))))
       ((> -y -x)
        (cond ((and (zerop (caddr c1)) (zerop (cadddr c2))) 0)
              (t (born (divide (differ (car c1) (cadr c2))
                               (plus (cadddr c2) (caddr c1))))))))))
;Computes the possibility of "c1" having "c2" ("c1" and "c2" are 4-tuple
;representations of distribution)

(defun born (x) (maxr 0 (differ x 1)))

(defun neccc (c1 c2)
  (cond ((and (zerop (caddr c1)) (zerop (caddr c2))
             (zerop (cadddr c1)) (zerop (cadddr c2)))
        (cond ((< (car c2) (car c1)) 0)
```

```
                       (((<= (cadr c2) (cadr c1)) 1)
                        (t 0)))
              ((and (zerop (cadddr c1)) (zerop (cadddr c2)))
               (cond (((< (cadr c1) (cadr c2)) 0)
                      (t (minr 1 (maxr 0 (divide (plus (caddr c1) (car c2)
                                                       (* -1 (car c1)))
                                             (plus (caddr c2) (caddr c1)))))))))
              ((and (zerop (caddr c1)) (zerop (caddr c2)))
               (cond (((< (car c2) (car c1)) 0)
                      (t (minr 1 (maxr 0 (divide (plus (cadddr c1) (cadr c1)
                                                       (* -1 (cadr c2)))
                                             (plus (cadddr c2) (cadddr c1)))))))))
              (t (minr 1 (minr (maxr 0 (divide (plus (caddr c1) (car c2)
                                                     (* -1 (car c1)))
                                           (plus (caddr c2) (caddr c1))))
                            (maxr 0 (divide (plus (cadddr c1) (cadr c1)
                                                  (* -1 (cadr c2)))
                                        (plus (cadddr c2) (cadddr c1)))))))))
;Computes the necessity of "c1" having "c2" ("c1" and "c2" are 4-tuple
;representations of distribution)


(defun crosref (regles)
   (cond ((null regles) nil)
         ((cr1rg (cadr (eval (car regles))))(crosref (cdr regles)))))
;For attaching to each attribute the list of rules concluding on it.
;"regles"= the set of rules analyzed one by one by "cr1rg"

(defun cr1rg (x)
(let ((u nil) (v nil))
   (cond ((null x) t)
         ((member (car regles) (get (caar x) 'crosref))(cr1rg (cdr x)))
         (t (putprop (caar x) (append (get (caar x) 'crosref)
                 (list (car regles))) 'crosref)(cr1rg (cdr x))))))
;Analysis of the consequent part (i.e."x") of the rule (car regles).

(defun  attrfait (bf)
(let ((x nil))
(cond ((null bf) nil)
      (t (putprop (setq x (car (eval (car bf))))
                  (append (get x 'pbf) (list (car bf))) 'pbf)
         (putprop (car bf) t 'exhausted)
         (attrfait (cdr bf))))))
;For attaching to each attribute the list of facts using it (useful for
;speeding the pattern-matching process).
;Also puts the property "exhausted" to each initial fact.


;Once the system is loaded and before querying the system the following
;has to be done.
(load "kbase")(load "database") ;load the knowledge and data bases.
(setq compt~ 0 comptf 100)
(crosref regles)  ;"regles" is the list of rule names.
(attrfait bf) ;"bf" is the list of fact names.



;****************************
;        KNOWLEDGE BASE


(setq r1 '(((special-skills applicant ok))
           ((quality applicant ok .7 0)))) ;sufficiency =.7, necessity =0

(setq r2 '(((iq-test applicant ~10-14) (french-test applicant ~6-9))
           ((test-scores applicant ~medium))))
(putprop 'r2 t 'floue) ;i.e.r2 has to be used with the generalized modus ponens
;The value components starting with ~ are imprecise (i.e.represented by
;possibility distributions)

(setq r3 '(((iq-test applicant ~14-20) (french-test applicant ~6-9))
           ((test-scores applicant ~good))))
(putprop 'r3 t 'floue) ;i.e.r3 has to be used with the generalized modus ponens

(setq r4 '(((iq-test applicant ~10-14) (french-test applicant ~9-20))
           ((test-scores applicant ~satisfactory))))
```

```
(putprop 'r4 t 'floue) ;i.e.r4 has to be used with the generalized modus ponens

(setq r5 '((((iq-test applicant ~14-20) (french-test applicant ~9-20))
            ((test-scores applicant ~excellent))))
(putprop 'r5 t 'floue) ;i.e.r5 has to be used with the generalized modus ponens

(setq r6 '((((work-experience applicant ok) (adaptability applicant ok))
            ((basic-skills applicant ok .7 .4))))

(setq r7 '((((basic-skills applicant ok))
            ((quality applicant ok 1 1))))

(setq r8 '((((manual-skills applicant ok))
            ((special-skills applicant ok .6 0))))

(setq r9 '((((sports-skills applicant ok))
            ((special-skills applicant ok .7 .5))))

(setq r10 '((((artistic-skills applicant ok))
            ((special-skills applicant ok .8 0))))

(setq r11 '((((education applicant ok) (test-scores applicant ~greater-medium))
            ((basic-skills applicant ok 1 1))))

(setq r12 '((((oral-test applicant ~6-9)
               (written-test applicant ~greater-medium))
            ((french-test applicant ~medium))))
(putprop 'r12 t 'floue) ;i.e r12 has to be used with the generalized modus ponens

(setq r13 '((((oral-test applicant ~greater-medium)
               (written-test applicant ~greater-medium))
            ((french-test applicant ~good))))
(putprop 'r13 t 'floue) ;i.e r13 has to be used with the generalized modus ponens

(setq r14 '((((oral-test applicant ~greater-medium)
               (written-test applicant ~6-9))
            ((french-test applicant ~satisfactory))))
(putprop 'r14 t 'floue) ;i.e r14 has to be used with the generalized modus ponens


(setq regles '(r1 r2 r3 r4 r5 r6 r7 r8 r9 r10 r11 r12 r13 r14)) ;set of rules


;Definition of the trapezoidal possibility distributions involved in the rules.
(setq ~greater-medium '(13 19 4 1))
(setq ~10-14 '(10 14 1 1))
(setq ~14-20 '(14 20 2 0))
(setq ~9-20 '(9 20 1 0))
(setq ~6-9 '(6 9 2 2))
(setq ~medium '(11 12 3 1))
(setq ~good '(15 16 2 1))
(setq ~satisfactory '(12.5 14 1.5 1))
(setq ~excellent '(17 20 2 0))




;*****************************
;           DATA BASE
;

(setq f1 '(education applicant OK))
(defprop f1 (.8 1) cert) ;necessity=.8 , possibility=1

(setq f2 '(iq-test applicant ~12))
(defprop f2 (1 1) cert)

(setq f3 '(oral-test applicant ~rather-good))
(defprop f3 (.9 1) cert)

(setq f4 '(written-test applicant ~5-7))
(defprop f4 (1 1) cert)

(setq f5 '(work-experience applicant OK))
(defprop f5 (1 1) cert)

(setq f6 '(adaptability applicant OK))
```

```
(defprop f6 (0 .6) cert) ;i.e f6 is rather false

(setq f7 '(artistic-skills applicant OK))
(defprop f7 (0 .2) cert) ;i.e f7 is rather false

(setq f8 '(sports-skills applicant OK))
(defprop f8 (1 1) cert)

(setq bf '(f1 f2 f3 f4 f5 f6 f7 f8)) ;set of initial facts


;Possibility distributions used in the initial facts
(setq ~rather-good '(14 15 1 1))
(setq ~5-7 '(5 7 1 1))
(setq ~12 '(12 12 2 1))
```

TRANSCRIPT OF A RUN AND COMMENTS

```
(apropos '(quality applicant ok)) ;query about the quality of an applicant.
certitude = ( 7.000000E-1 1) ;necessity-possibility pair answered by the system.
= ( 7.000000E-1 1)

bf ;actual state of the data base : five new facts added.
= (f1 f2 f3 f4 f5 f6 f7 f8 f101 f102 f103 f104 f105)

f101
= (special-skills applicant ok)
(get 'f101 'cert) ;to get the necessity-possibility pair attached to f101
= ( 7.000000E-1  1.000000E0)
(get 'f101 'provient) ;to get how and with what uncertainty f101 has been deduced
= ((r9 7.000000E-1  1.000000E0))

f102
= (quality applicant ok)
(get 'f102 'cert)
= ( 7.000000E-1 1)
(get 'f102 'provient)
= ((r1 7.000000E-1 1) (r7 3.939394E-1 1))

f103
= (basic-skills applicant ok)
(get 'f103 'cert)
= ( 3.939394E-1 1)
(get 'f103 'provient)
= ((r6 0.0 6.000000E-1) (r11 6.363636E-1 1))
;r6 and r11 provide rather conflicting conclusions but r11 is slightly more
;affirmative. Thus the combination of these conclusions yields that f103 is
;rather true

f104
= (french-test applicant ~2)
(get 'f104 'cert)
= ( 9.000000E-1 1)
(get 'f104 'provient) ;returns also the fuzzy values deduced by each rules
= ((r12 ~indetermine 1 1) (r13 ~indetermine 1 1) (r14 ~1 9.000000E-1 1))
~2 ;possibility distribution obtained by combining values produced by r12 r13 r14
= ( 1.175000E1  1.450000E1  8.333334E-1  5.555556E-1)
~1 ;possibility distribution produced by r14
= ( 1.175000E1  1.450000E1  8.333334E-1  5.555556E-1)
;f104 is obtained via the combination of conclusions deduced from r12, r13, r14,
;each applied through the generalized modus ponens technique. Actually, r12 and
;r13 provide conclusions that are completely indeterminate. Therefore the value
;component of f104 is exactly the one produced by r14.

f105
= (test-scores applicant ~4)
(get 'f105 'cert)
= ( 9.000000E-1 1)
(get 'f105 'provient)
= ((r2 ~indetermine 1 1) (r3 ~indetermine 1 1) (r4 ~3 9.000000E-1 1)
   (r5 ~indetermine 1 1))
;f105 is also deduced via the generalized modus ponens technique.
~4 ;possibility distribution obtained after combination
```

```
= ( 1.250000E1 14  1.500000E0 1)
~3
= ( 1.250000E1 14  1.500000E0 1)
;Here again only one rule (i.e.r4) gives a nontrivial conclusion about the
;test scores of the applicant.



!   2. DEDUCTION FROM UNCERTAIN FACTS AND RULES
!===============================================================================
! Constants :
!------------
oui=1
non=0
maxfait=20
maxnoeud=50
STRSIZ 80
!Reading of the rules :
!---------------------
  read nbregles
  dim premisse$(nbregles),concl$(nbregles)
  dim necessity(nbregles),sufficiency(nbregles)
  for I=1 to nbregles
    read premisse$(I),concl$(I),necessity(I),sufficiency(I)
  next I

!Reading of the facts ('faits') :
!-------------------------------
  dim fait$(maxfait),necfait(maxfait),posfait(maxfait),
  read nbfaits
  for I=1 to nbfaits
    read fait$(I),necfait(I),posfait(I)
  next I

!Reservations for the evaluation tree :
!-------------------------------------
  dim noeud$(maxnoeud)                    ! evaluation tree
  dim pos(maxnoeud)                       ! possibility of the node ('noeud')
  dim nec(maxnoeud)                       ! necessity of the node
  dim determine(maxnoeud)                 ! 'oui' if pos<>1 and nec<>0, 'non' otherwise
  dim pere(maxnoeud)                      ! pointer to the father node
  dim nbfils(maxnoeud)                    ! number of sons ('fils') of a node
  dim premfils(maxnoeud)                  ! index of the first son
  dim nbflsexam(maxnoeud)                 ! number of sons already considered
  dim nbrfilsdetermine(maxnoeud)          ! number of sons already evaluated

!Control of the program :
!-----------------------
MENU:
  print tab(-1,0)
CONTROL :
  print tab(10);"1. consulting of the fact base"
  print tab(10);"2. attempt to evaluate a fact"
  print tab(10);"3. end "
  print
  print "choice of the option"
ANSWER :
  input R
  if R<1 or R>3 then goto ANSWER
  on R goto CONSULT'FACT,DEM'FACT,OUTPUT


CONSULT'FACT :
  for I=1 to nbfaits
    print fait$(I);tab(40);necfait(I);tab(60);posfait(I)
  next I
  goto CONTROL

DEM'FACT :
  INPUT "FACT TO EVALUATE : ",faidatet$
  gosub INIT'DEM
  T=1
```

```
        T0=1
        noeud$(T0)=faitadet$
        gosub FACT'EVALUATION
        print determine(T0),nec(T0),pos(T0)
        GOTO CONTROLE

OUTPUT :
        end
```

```
!                           INFERENCE ENGINE
!=====================================================================
```

```
! --------------------------------------------------------------------
!                    SUBROUTINE FACT'EVALUATION
!                    -------------------------------
!
!   This subroutine computes estimates of the possibility ('pos') and of the ne-
!   cessity ('nec') that a fact ('fait'), corresponding to a node ('noeud') T0,
!   is true. The result is obtained either by direct consultation of the know
!   edge base or by the use of one or several rules concluding on this fact.
!   When this is done the value 'oui' ('yes') is assigned to 'determine (T0)'.
!
! --------------------------------------------------------------------

FACT'EVALUATION :

    ! If T0 corresponds to a known fact then its possibility and necessity are already known
        for I=1 to nbfaits
            if noeud$(T0)=fait$(I) then &
                pos(T0)=posfait(I):&
                nec(T0)=necfait(I):&
                determine(T0)=oui:&
                I=nbfaits
        next I

        if determine(T0)=oui then return

    ! T0 may then corresponds to the conclusion of one rule ('règle')
        for I=1 to nbregles
            if noeud$(T0)=concl$(I) then &
                T=T+1:&
                noeud$(T)=$TR(I):&
                pere(T)=T0:&                            ! 'pere' and 'fils' mean
                nbfils(T0)=nbfils(T0)+1:&               ! 'father' and 'son' respectively
                if nbfils(T0)=1 then premfils(T0)=T
        next I

    ! If T0 is not the conclusion of any rule then
    ! the fact cannot be evaluated.
        if nbfils(T0)=0 then &
            nec(T0)=0:&
            pos(T0)=1:&
            determine(T0)=non:&
            return

    ! If it is not the case, the program looks for a rule to apply
        for nbflsexam(T0)=1 to nbfils(T0)
            T0=premfils(T0)+nbflsexam(T0)-1
            gosub APPLY'RULE
        next nbflsexam(T0)

    ! If no rule is applicable then the fact cannot be evaluated
        if nbrfilsdetermine(T0)=0 then &
            pos(T0)=1:&
            nec(T0)=0:&
            determine(T0)=non:&
            return

    ! If it is not the case there is one or several rules which are applicable
    ! and then we may have to combine their conclusions
        p=1
        n=0
        for I=premfils(T0) to premfils(T0)+nbfils(T0)-1
            if determine(I)=oui then &
                p=p min pos(I):&
```

```
        n=n max nec(I)
  next I

  !normalization
  pos(T0)=p/((1-n) max p)
  nec(T0)=1-(1-n)/((1-n) max p)
  if pos(T0)<>1 or nec(T0)<>0 then determine(T0)=oui
return
```

```
!                           SUBROUTINE APPLY'RULE
!                           -------------------------
! This subroutine tries to apply a rule whose number is in noeud$(T0). First
! the condition (or premise) part of the rule is evaluated. Then the uncer-
! tainty is propagated along the rule.
!
!-------------------------------------------------------------------------------
APPLY'RULE
  !A leaf is created for each elementary premise to evaluate
      CH$=premisse$(VAL(noeud$(T0)))

  PREMISSE:
      gosub HEAD'TAIL
      T=T+1
      noeud$(T)=HEAD$
      pere(T)=T0
      nbfils(T0)=nbfils(T0)+1
      if nbfils(T0)=1 then premfils(T0)=T
      if TAIL$:&
        CH$=TAIL$:&
        goto PREMISSE

  !Then it is checked if each premise is evaluated
      for nbflsexam(T0)=1 to nbfils(T0)
        T0=premfils(T0)+nbflsexam(T0)-1
        gosub DETERMINE'FACT
        if determine(T0)=oui then &
          nbrfilsdetermine(pere(T0))=nbrfilsdetermine(pere(T0))+1
        T0=pere(T0)
      next nbflsexam(T0)

  !If no premise can be evaluated then the rule is inapplicable
      if nbrfilsdetermine(T0)=0 then &
        pos(T0)=1:&
        nec(T0)=0:&
        determine(T0)=non:&
        T0=pere(T0):&
        return

  !If at least one premise can be evaluated, the maximum amount of
  !uncertainty is estimated
      n=1
      p=1
      for I=1 to nbfils(T0)
        n=n min nec(premfils(T0)+I-1)
        p=p min pos(premfils(T0)+I-1)
      next I

  !Propagation of the uncertainty
      pos(T0)=p max (1-necessity(VAL(noeud$(T0))))
      nec(T0)=n min sufficiency(VAL(noeud$(T0)))

  !Back to an upper level in the tree
      if nec(T0)<>0 or pos(T0)<>1 then determine(T0)=oui :&
                nbrfilsdetermine(pere(T0))=nbrfilsdetermine(pere(T0))+1
      T0=pere(T0)
return
```

```
!                           SUBROUTINE HEAD/TAIL
!                           ----------------------- - - -
! From a list CH$ of literal chains separated by points, this subroutine
! yields the first chain in HEAD$ and the remaining of the list in TAIL$.
!
!-------------------------------------------------------------------------------
```

```
HEAD'TAIL :

  I1=INSTR(1,CH$,".")
  if  I1=0  then  HEAD$=CH$:TAIL$=""&
            else  HEAD$=CH$[1,I1-1]:TAIL$=CH$[I1+1,LEN(CH$)]
  return

!---------------------------------------------------------------------------
!                        SUBROUTINE INIT'DEM
!                        ---------------------
!   This subroutine reinitializes all the tables used for representing the
!   evaluation tree.
!---------------------------------------------------------------------------
INIT'DEM:

    print "initialization"
    for I=1 to T
      nceud$(I)=0
      pos(I)=0
      nec(I)=0
      pere(I)=0
      determine(I)=0
      nbfils(I)=0
      premfils(I)=0
      nbflsexam(I)=0
      nbrfilsdetermine(I)=0
    next I
    return

++INCLUDE SELECT.BSI

rem Rule base :
rem ----------
data 17
data "Good' IQ level. Human'qualities", "Good opinion",1,1
data "Good'professional'education,"Good' IQ'level", .2, .9
data "General'background", "Good'IQ'level", .9, .8
data "School. Visits", "Good'professional'education", .2, 1
data "Education", "General'background", .6, 1
data "Open'mind", "General'background "t", .3, .8
data "Responsible", "Human'qualities", .4, 1
data "Efficient", "Human'qualities", .5, .5
data "Friendly", "Human'qualities", .5, 1
data "Decider", "Responsible", .7, 1
data "Balance. Judgement. Initiative" ; "Responsible",  .8, 1
data "Method. Intuition", "Efficient", .3, .9
data "Analytical'mind", "Efficient", .4.,.1
data "Talkative. Patient", "Friendly", 1, 1
data "Sensitiveness. Humor", "Friendly", .5, 1
data "Nice'looking", "Friendly", .9, .6
data "Fast'minded" . "Eager'at'understanding", "Initiative", 0.6, 1

rem Facts :
rem -------
data 10
data "School", 1,1
data "Education", .5,1
data "Judgement",.8,1
data "Sensitiveness",1,1
data "Humor",.8,1
data "Good'quality'of'speech",0,.2
data "Method",1,1
data "Intuition",.2,1
data "Fast'minded",1,1
data "Eager'at'understanding', .7,1
```

References

1. ADLASSNIG, K. P., and KOLARZ, G. (1982). CADIAG-2: Computer-assisted medical diagnosis using fuzzy subsets. In *Approximate Reasoning in Decision Analysis* (M. M. Gupta and E. Sanchez, eds.). North-Holland, Amsterdam, pp. 219–247.

2. BALDWIN, J. F. (1979). A new approach to approximate reasoning using a fuzzy logic. *Fuzzy Sets Syst.*, **2**, 309–325.
3. BALDWIN, J. F., and PILSWORTH, B. W. (1980). Axiomatic approach to implication for approximate reasoning with fuzzy logic. *Fuzzy Sets Syst.*, **3**, 193–219.
4. BELLMAN, R. E., and ZADEH, L. A. (1977). Local and fuzzy logics. In *Modern Uses of Multiple-Valued Logic* (J. M. Dunn and G. Epstein, eds.). D. Reidel, Dordrecht, pp. 103–165.
5. CAYROL, M., FARRENY, H., and PRADE, H. (1982). Fuzzy pattern matching. *Kybernetes*, **11**, 103–116.
6. DEMPSTER, A. P. (1967). Upper and lower probabilities induced by a multivalued mapping. *Ann. Math. Stat.*, **38**, 325–339.
7. DUBOIS, D., and PRADE, H. (1979). Operations in a fuzzy-valued logic. *Inf. Control*, **43**(2), 224–240.
8. DUBOIS, D., and PRADE, H. (1982). A class of fuzzy measures based on triangular norms—A general framework for the combination of uncertain information. *Int. J. Gen. Syst.*, **8**(1), 43–61.
9. DUBOIS, D., and PRADE, H. (1982). Degree of truth and truth-functionality. *Proc. 2nd World Conf. on Maths at the Service of Man*, Las Palmas, Spain, June 28–July 3, 1982, pp. 262–265.
10. DUBOIS, D., and PRADE, H. (1982). On several representations of an uncertain body of evidence. In *Fuzzy Information and Decision Processes* (M. M. Gupta and E. Sanchez, eds.), North-Holland, Amsterdam, pp. 167–181.
11. DUBOIS, D., and PRADE, H. (1983). On distances between fuzzy points and their use for plausible reasoning. *Proc. IEEE Int. Conf. on Cybernetics and Society,* Bombay–New Dehli, Dec. 30, 1983–Jan. 7, 1984, pp. 300–303.
12. DUBOIS, D., and PRADE, H. (1984). Fuzzy logics and the generalized modus ponens revisited. *Cybern. Syst.*, **15**, 293–331.
13. DUBOIS, D., and PRADE, H. (1985). The generalized modus ponens under sup–min composition. A theoretical study. In *Approximate Reasoning in Expert Systems* (M. M. Gupta, A. Kandel, W. Bandler, and J. B. Kiszka, eds.), North-Holland, Amsterdam, pp. 217–232.
14. DUBOIS, D., and PRADE, H. (1987). The management of uncertainty in fuzzy expert systems and some applications. In *The Analysis of Fuzzy Information*, Volume 2: Artificial Intelligence and Decision Systems (J. C. Bezdek, ed.). CRC Press, Boca Raton, Florida, pp. 39–58.
15. DUDA, R., GASCHNIG, J., and HART, P. (1981). Model design in the PROSPECTOR consultant system for mineral exploration. *Expert Systems in the Micro-Electronic Age* (D. Michie, ed.). Edingurgh University Press, pp. 153–167.
16. ERNST, C. (1982). Le modèle de raisonnement approché du système MANAGER. BUSEFAL, Report No. 9, L.S.I., University P. Sabatier, Toulose, pp. 93–99.
17. FRIEDMAN, L. (1981). Extended plausible inference. *Proc. 7th Int. Joint. Conf. Artificial Intelligence.* Vancouver, August 1981, pp. 487–495.
18. GAINES, B. R. (1976). Foundations of fuzzy reasoning. *Int. J. Man-Machine Stud.*, **8**, 623–668.
19. GOGUEN, J. A. (1969). The logic of inexact concepts. *Synthese*, **19**, 325–373.
20. ISHIZUKA, M. (1983). Inference methods based on extended Dempster and Shafer's theory for problems with uncertainty/fuzziness. *New Generation Comp.*, **1**, 159–168.
21. ISHIZUKA, M., FU, K. S., and YAO, J. T. P. (1981). Inexact inference for rule-based damage assessment of existing structures. *Proc. 7th Int. Joint Conf. on Artificial Intelligence.* Vancouver, pp. 837–842.

22. ISHIZUKA, M., FU, K. S., and YAO, J. T. P. (1982). Inference procedures with uncertainty for problem reduction method. *Inf. Sci.,* **28,** 179–206.
23. KAYSER, D. (1979). Vers une modélisation du raisonnement "approximatif." Proc. of the Colloquium *Représentation des Connaissances et Raisonnement dans les Sciences de l'Homme* (M. Borillo, ed.), Saint-Maximin, September 1979, Published by INRIA, pp. 440–457.
24. LESMO, L., SAITTA, L., and TORASSO, P. (1983). Fuzzy production rules: A learning methodology. In *Advances in Fuzzy Set Theory and Applications,* (P. P. Wang, ed.). Plenum Press, New York, pp. 181–198.
25. MAMDANI, E. H. (1977). Application of fuzzy logic to approximate reasoning using linguistic systems. *IEEE Trans. Comput,* **C-26,** 1182–1191.
26. DUBOIS, D., MARTIN-CLOUAIRE, R., and PRADE, H. (1988). Practical computing in fuzzy logics. In *Fuzzy Computing* (M. M. Gupta and T. Yamakawa, eds.), North-Holland, Amsterdam, to be published.
27. MARTIN-CLOUAIRE, R., and PRADE, H. (1985). On the problems of representation and propagation of uncertainty in expert systems. *Int. J. Man-Machine Stud.,* **22,** 251–264.
28. LEBAILLY, J., MARTIN-CLOUAIRE, R., and PRADE, H. (1987). Use of fuzzy logic in a rule-based system in petroleum geology. In *Approximate Reasoning in Intelligent Systems, Decision and Control* (E. Sanchez and L. A. Zadeh, eds.), Pergamon, Oxford, pp. 125–144.
29. MIZUMOTO, M., FUKAMI, S., and TANAKA, K. (1979). Fuzzy conditional inferences and fuzzy inferences with fuzzy quantifiers. *Proc. 6th Int. Joint Conf. on Artificial Intelligence,* Tokyo, pp. 589–591.
30. MIZUMOTO, M., and ZIMMERMANN, H. J. (1982). Comparison of fuzzy reasoning methods. *Fuzzy Sets Syst.,* **8,** 253–283.
31. POLYA, G. (1954). *Mathematics and Plausible Reasoning.* Vol. II: *Patterns of Plausible Inference.* Princeton University Press, Princeton, New Jersey, 2nd edition 1968.
32. PRADE, H. (1982). Modèles mathématiques de l'imprécis et de l'incertain en vue d'applications au raisonnement naturel. Thesis University of Toulose III, June 1982 (358 pp.).
33. PRADE, H. (1983). A synthetic view of approximate reasoning techniques. *Proc. 8th Int. Joint Conf. Artificial Intelligence,* Karlsruhe, August 1983, pp. 130–136.
34. PRADE, H. (1983). Data bases with fuzzy information and approximate reasoning in expert systems. *Proc. IFAC Int. Symp. Artificial Intelligence.* Leningrad, USSR, October 4–6, 1983, pp. 113–120.
35. PRADE, H. (1984). Modèles de raisonnement approché pour les systèmes experts. *Proc 4ème Congrès AFCET "Reconnaissance des Formes & Intelligence Artificielle,* Paris, January 25–27, 1984, pp. 355–373.
36. PRADE, H. (1985). A computational approach to approximate and plausible reasoning, with applications to expert systems. *IEEE Trans. Pattern Anal. Machine Intell.,* **7,** 260–283. Corrections in **7,** 747–748.
37. RESCHER, N. (1969). *Many-Valued Logic.* McGraw-Hill, New York.
38. SANCHEZ, E. (1976). Resolution of composite fuzzy relation equations. *Inf. Control,* **30,** 38–48.
39. SHAFER, G. (1976). *A Mathematical Theory of Evidence.* Princeton University Press, Princeton, New Jersey.
40. SHORTLIFFE, E. H., and BUCHANAN, B. G. (1975). A model of inexact reasoning in medicine. *Math. Biosci.,* **23,** 351–379.
41. SOULA, G., and SANCHEZ, E. (1982). Soft deduction rules in medical diagnosis

processes. In *Approximate Reasoning in Decision Analysis,* (M. M. Gupta and E. Sanchez, eds.), North-Holland, Amsterdam, pp. 77–88.

42. SOULA, G., VIALETTES, B., and SAN MARCO, J. L. (1983). PROTIS, a fuzzy deduction-rule system: Application to the treatment of diabetes. *Proc. Medinfo 83,* (Van Bemmel, Ball, and Wigertz, eds.), IFIP-INIA, Amsterdam, pp. 553–536.

43. SUPPES, P. (1966). Probabilistic inference and the concept of total evidence. In *Aspects of Inductive Logic* (J. Hintikka and P. Suppes, eds.), North-Holland, Amsterdam, pp. 49–65.

44. TONG, R. M., SHAPIRO, D. G., DEAN, J. S., and McCUNE, B. P. (1983). A comparison of uncertainty calculi in an expert system for information retrieval. *Proc. 8th Int. Joint Conf. Artificial Intelligence,* Karlsruhe, August 1983, pp. 194–197.

45. TRILLAS, E., and VALVERDE, L. (1981). On some functionally expressible implications for fuzzy set theory. *Proc. 3rd Int. Seminar on Fuzzy Set Theory* (E. P. Klement, ed.), J. Kepler Univ., Linz, Austria, September 7–12, 1981, pp. 173–190.

46. TSUKAMOTO, Y. (1979). An approach to fuzzy reasoning method. In *Advances in Fuzzy Set Theory and Applications* (M. M. Gupta, R. K. Ragade, and R. R. Yager, eds.). North-Holland, Amsterdam, pp. 137–149.

47. WEBER, S. (1983). A general concept of fuzzy connectives, negations and implications based on t-norms and t-co-norms. *Fuzzy Sets and Syst.,* **11,** 115–134.

48. WHALEN, T., and SCHOTT, B. (1983). Issues in fuzzy production systems. *Int. J. Man-Machine Stud.,* **19,** 57–71.

49. YAGER, R. R. (1980). An approach to inference in approximate reasoning. *Int. J. Man-Machine Stud.,* **13,** 323–338.

50. ZADEH, L. A. Outline of a new approach to the analysis of complex systems and decision processes. *IEEE Trans. Syst., Man Cybernet.,* **3,** 28–44.

51. ZADEH, L. A. (1978). PRUF-A meaning representation language for natural languages. *Int. J. Man-Machine Stud.,* **10**(4), 395–460.

52. ZADEH, L. A. (1979). A theory of approximate reasoning. *Machine Intelligence, Vol. 9,* (J. E. Hayes, D. Mitchie, and L. I. Mikulich, eds.). Elsevier, New York, pp. 149–194.

53. ZADEH, L. A. (1983). The role of fuzzy logic in the management of uncertainty in expert systems. *Fuzzy Sets Syst.,* **11**(3), 199–228.

54. ZADEH, L. A. (1984). Review of " A Mathematical Theory of Evidence," by G. Shafer. *The AI Magazine,* Fall 1984, 81–83.

55. APPELBAUM, L., RUSPINI, E. H. (1985). ARIES: An approximate reasoning inference engine. In *Approximate Reasoning in Expert Systems,* (M. M. Gupta, A. Kandel, W. Bandler, and J. B. Kiszka, eds.). North-Holland, Amsterdam, pp. 745–765.

56. BUCKLEY, J., SILVER, W., and TUCKER, D. (1986). FLOPS: A fuzzy expert system: applications and perspectives. In *Fuzzy Logic in Knowledge Engineering* (C. V. Negoita and H. Prade, eds.). Verlag TÜV Rheinland, Cologne.

57. CHATALIC, P., DUBOIS, D., and PRADE, H. An approach to approximate reasoning based on Dempster rule of combination. *Proc. 8th IASTED Inter. Symp. Robotics and Artificial Intelligence,* Toulouse, France, June 18–20, pp. 333–343.

58. DUBOIS, D., and PRADE, H. (1984). A theorem on implication functions defined from triangular norms. *Stochastica,* **VIII,** 267–279.

59. DUBOIS, D., and PRADE, H. (1985). Combination and propagation of uncertainty with belief functions. A reexamination. *Proc. 9th Inter. Joint Conf. Artificial Intelligence,* Los Angeles, pp. 111–113.

60. DUBOIS, D., and PRADE, H. (1986). Possibilistic inference under matrix form. In *Fuzzy Logic in Knowledge Engineering* (C. V. Negoita, and H. Prade, eds.), Verlag TÜV Rheinland, Cologne, pp. 112–126.

61. FARRENY, H., and PRADE, H. (1986). Default and inexact reasoning with possibility degrees. *IEEE Trans. Syst., Man. Cybernet.*, **16**, 270–276.
62. FARRENY, H., PRADE, H., and WYSS, E. (1986). Approximate reasoning in a rule-based expert system using possibility theory: A case study. *Proc. 10th IFIP World Computer Cong.*, Dublin, Ireland, September, 1–5.
63. GARVEY, T. D., LOWRANCE, J. D., and FISCHLER, M. A. (1985). An inference technique for integrating knowledge from disparate sources. *Proc. 7th Inter. Joint Conf. Artificial Intelligence*, Vancouver, August 1981, pp. 319–325.
64. HISDAL, E. (1978). Conditional possibilities. Independence and non-interactivity. *Fuzzy Sets Syst.*, **1**, 283–297.
65. MARTIN-CLOUAIRE, R., and PRADE, H. SPII-1: A simple inference engine for accommodating both imprecision and uncertainty. In *Computer-Assisted Decision Making*, North-Holland, Amsterdam, pp. 117–131.
66. NEGOITA, C. V., and PRADE, H. (eds.). (1986). *Fuzzy Logic in Knowledge Engineering*. Verlag TÜV Rheinland, Cologne.
67. FIESCHI, M. (1984). *Intelligence artificielle en médecine. Des systèmes experts. Méthode + Programmes series*. Masson, Paris.
68. WEBER, S. (1984). ⊥ -decomposable measures and integrals for Archimedean *t*-conorms ⊥. *J. Math. Anal. Appl.*, **101**, 114–138.
69. DUBOIS, D., and PRADE, H. (1988). On the combination of uncertain or imprecise pieces of information in rule-based systems. *Inter. J. of Approximate Reasoning*, to be published.
70. PEDRYCZ, W. (1985). On generalized fuzzy relational equations and their applications. *J. Math. Anal. Appl.*, **107**, 520–536.
71. SUGENO, M. (ed). (1985). *Industrial Applications of Fuzzy Control*. North-Holland, Amsterdam.
72. BUISSON, J. C., FARRENY, H., PRADE, H., TURNIN, M. C., TAUBER J. P., and BAYARD, F. (1987). TOULMED, an inference engine which deals with imprecise and uncertain aspects of medical knowledge. *Proc. AIME 87*, (J. Fox, M. Fieschi, and R. Engelbrecht, eds.), *Lecture Notes in Medical Informatics*, Vol. 33, Springer Verlag, Berlin, 123–140.
73. SANCHEZ, E., ZADEH, L. A. (eds.). (1987). *Approximate Reasoning in Intelligent Systems, Decision and Control*. Pergamon, Oxford.

5

Heuristic Search in an Imprecise Environment, and Fuzzy Programming

The aim of artificial intelligence is to mimic, by machine, operations that the human mind can readily achieve, though we may not know exactly how. For example, understanding messages, making plans of action, analyzing situations, adapting a general mode of behavior to particular circumstances .... In all these activities the human being may have to take account of imprecise and uncertain information. However, this aspect of human intelligence has hitherto been relatively little studied in artificial intelligence.

In the preceding chapter we introduced inference mechanisms that facilitate use of such imprecise and uncertain information. In the first part of this chapter we shall be concerned with some techniques in which solving a problem is viewed as a search for a sequence of elementary actions by means of an evaluation function. Conversely, in the second part we shall suppose that we already have available a sequence of elementary actions, and that it is a question of carrying them out in a practical context.

In the first case, the imprecision is associated with the operands of the evaluation function that governs the search. In the second case, the imprecision may appear equally in the specification of the plan of action and in the perception of the environment in the course of executing the proposed plan.

The chapter does not presume to reflect all the research that is aimed at applying possibility theory to artificial intelligence. Topics that we shall not deal with here include concept acquisition (see [2]).

5.1. Heuristic Search in an Imprecise Environment

Search techniques on graphs, and especially on rooted trees, are extensively employed in operational research (see Gondran and Minoux [10]) as well as in artificial intelligence (Nilsson [13, 14], Winston [18], Pearl [15]).

In this section we consider the A^* family of algorithms (Nilsson [14]) whose basis is described below. As an illustrative example, a procedure from this family will be applied to the traveling salesman problem, in the first place with precise evaluations of the costs, then with interval representations, and finally with the cost evaluations represented as fuzzy intervals.

5.1.1. A and A^* Algorithms

The solution of a problem will be viewed as a search for a path through a graph, from a fixed vertex called the initial state to a vertex in a preassigned subset called the set of terminal states. The graph, which will be called the state space, is supposed to be only potentially known, in the sense that we only have available, apart from the initial and the final states, a set of operators that can generate the successors of a given state.

Application of such an operator to a given state ("father") is represented by a directed edge from the "father" to each engendered state ("son"). Each edge is associated with a cost, which we shall take to be positive. The aim of the search is to construct a path of minimum cost from the initial state to a terminal state, where the cost of a path will be evaluated as the sum of the costs of its edges. The search procedure consists of progressively engendering the progeny of a given initial state until a terminal state appears. The graph that emerges at any given stage of the search is called the "search graph"; when a terminal state has not yet appeared in the search graph there is the problem of choosing a state to which to apply the operators in order to engender new states ("developing a state"). This choice is guided by an "evaluation function," which, for each state n whose development is being considered, evaluates to a positive number $f(n)$ given by the sum

$$f(n) = g(n) + h(n) \tag{5.1}$$

The function g gives the cost of the least-cost path known at the current stage of the algorithm, between the initial state and the state n; the function h is a cost depending only on n, which we shall normally interpret as an estimate of the least cost of a path from n to some one of the terminal states. The state selected for development will be one of those for which the evaluation function f is least.

The algorithm tests whether a state is terminal, not when the state first appears in the search graph, but only when this state is chosen for development.

If the search graph is finite, and the heuristic term $h(n)$ is a lower bound for the minimum cost of a path between the state n and any one of the terminal states, the algorithm is of type A^* and it is then guaranteed that (1) the algorithm terminates; (2) at termination, either it yields a path to a terminal node when there is one, or it determines that there is no such path; and (3) when such a path to a terminal state is found, it is a least-cost path.

More generally, if it is not guaranteed that $h(n)$ is such a lower bound, the algorithm is only a type A algorithm (not type A^*), and only the properties (1) and (2) above are guaranteed. These properties still hold, with the support of a few supplementary hypotheses, in the case of infinite search graphs [15]. In what follows, an example of a type A^* algorithm will be described in the context of the so-called "traveling salesman" problem. It will then be modified for the case where the costs associated with the edges are only known to within classical intervals, and finally for the case of fuzzy intervals.

5.1.2. The Classical Traveling Salesman Problem (Reminder)

Let $G = (\mathscr{S}, \mathscr{A})$ be an arbitrary graph; the set $\mathscr{A}$ of edges represents road links between a set of towns represented by the elements of $\mathscr{S}$ (vertices). To each edge (S_i, S_j) corresponds a positive cost C_{ij}. The problem is to find, if it exists, a circuit that passes once and once only through each town and that has the least cost.

This problem can be solved by the A^* algorithm described below. The initial state is an arbitrary vertex S_0 of $\mathscr{S}$. The nonterminal intermediate states are sequences of vertices, all different, starting with S_0. The terminal states are sequences of vertices of which the first is S_0 and the remainder form a permutation of all the vertices of $\mathscr{S}$, ending with S_0. The state-space, not to be confused with the graph G itself, is formed in the following way: There is an edge from the state $n = (S_0, S_i, \ldots, S_j)$ to the state $n' = (S_0, S_i, \ldots, S_j, S_k)$ whenever the edge (S_j, S_k) exists in $\mathscr{A}$ and S_k does not belong to the subsequence $(S_i, \ldots, S_j)$. The term $g(n')$, for $n' = (S_0, S_i, \ldots, S_j, S_k)$, is the sum of the costs $C_{0,i} + \cdots + C_{j,k}$. The term $h(n')$ is calculated as follows: for each vertex not in the subsequence $S_i, \ldots, S_j, S_k$ we take the least cost of an edge meeting this vertex; $h(n')$ is the sum of these costs. This quantity

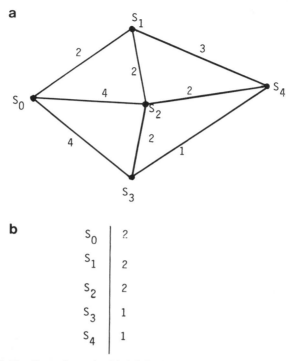

a

b

S_0	?
S_1	2
S_2	2
S_3	1
S_4	1

Figure 5.1. (a) Nondirected graph. (b) Minimal cost of an arc arriving at vertex S_i. (c) Search graph.

$h(n')$ is indeed a lower bound in the sense required by the definition of class A^* algorithms. Calculating f in this way, we obtain the path with the least cost.

Consider the example shown in Figure 5.1a. In Figure 5.1b we have calculated the minimum costs of the edges arriving at each vertex. Figure 5.1c shows the search graph obtained at the moment when the algorithm terminates; the terminal state thus discovered is $S_0 S_1 S_2 S_4 S_3 S_0$, whose cost is 11. Note that in the case where valuations are equal, we choose to develop those of greatest depth in the search graph. The default rule is that of the "leftmost."

Beside each state n in Figure 5.1c is marked the value of the valuation function $f(n)$, in the form $g(n) + h(n)$. We have also marked, in square frames, numbers representing the order of development of the different states. Thus, the development of $S_0 S_1 S_2 S_3 S_4$ was considered at the fifth stage (but in this case, no successor state can be found). The development of the states $S_0 S_2 S_3$, $S_0 S_2 S_4$, or $S_0 S_3$ might also give rise to

c

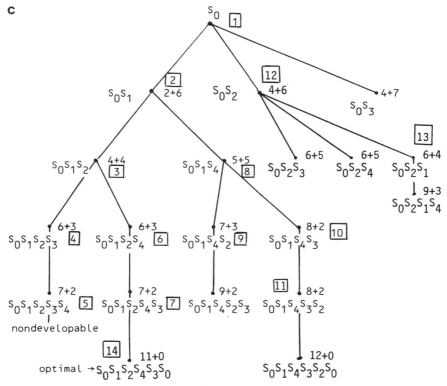

Figure 5.1. (*continued*).

optimal solutions. The reader may, however, verify that, in the example, it is not so.

5.1.3. Heuristic Search with Imprecise Evaluations

We now suppose that, by reason of incomplete knowledge of the conditions of the problem to be solved, the costs making up the evaluation function f are only available as intervals. Then the evaluations $f(n)$ are obtained by applying the rules of interval analysis (see Chapter 2). In order to adapt class A or A^* algorithms to the case where the evaluations are available in the form of intervals, we have to answer three questions:

1. What strategy should be used to choose a state for development?
2. What will the stopping rule of the procedure be?
3. What result will be obtained with such a rule?

The choice of state to be developed comes down to a problem of comparing intervals $[\underline{f}(n_j), \bar{f}(n_j)]$, $j = 1, \ldots, p$, where $n_1, \ldots, n_p$ are the candidate states for development at the current stage. We seek the state with the "smallest evaluation." When $p = 2$, we are led to compare the relative positions of the two intervals. Thus we obtain four possible classification criteria which indicate whether n_1, rather than n_2, is the state to develop:

C1: $\underline{f}(n_2) \geqslant \bar{f}(n_1)$
C2: $\underline{f}(n_2) \geqslant \underline{f}(n_1)$
C3: $\bar{f}(n_2) \geqslant \bar{f}(n_1)$
C4: $\bar{f}(n_2) \geqslant \underline{f}(n_1)$

If criterion C1 is satisfied, then we can be certain that the evaluation of n_1 is better than that of n_2, despite the imprecision. On the other hand, if criterion C4 is satisfied, all we know is that there remains a possibility that n_1 may have the smaller evaluation. When criteria C2 and C3 are simultaneously satisfied, we can write

$$C23: \quad \widetilde{\min} \left([\underline{f}(n_1), \bar{f}(n_1)], [\underline{f}(n_2), \bar{f}(n_2)] \right) = [\underline{f}(n_1), \bar{f}(n_1)]$$

where $\widetilde{\min}$ is the minimum operator applied to intervals by virtue of the extension principle (see Chapter 2). With the $\widetilde{\min}$ operator we can define a selection criterion C23 which is intuitively satisfying, and less stringent than C1. The selection criteria can be ordered according to their strength:

$$C1 \Rightarrow C23 \left\{ \begin{array}{l} \Rightarrow C2 \Rightarrow C4 \\ \Rightarrow C3 \Rightarrow C4 \end{array} \right.$$

In practice, we would suggest selecting the vertex to be developed by applying the criteria in the order indicated by the above implications. C2 and C3 are the only criteria not naturally ordered: we would choose C2 for preference if we thought that the smallest values of $f(n_1)$ and $f(n_2)$ were more plausible than the largest values. In the general case, when there are $p > 2$ states which can be developed, we are led to compare the evaluation of each n_j with

$$\widetilde{\min}_{k \neq j} [\underline{f}(n_k), \bar{f}(n_k)] = \left[\min_{k \neq j} \underline{f}(n_k), \min_{k \neq j} \bar{f}(n_k) \right]$$

that is, with the least of the other evaluations, using the criteria in the order suggested above.

When a state has been recognized as terminal according to a class A algorithm, one might wish to extend the search, when the criterion which led to its selection is judged to be undiscriminating. However, we may note that there does not necessarily exist an optimal state with respect to the criteria C1 or C23, when the evaluations are imprecise, even if an optimal state had been determined by the algorithm A^* in the problem with precise evaluations.

Thus if we intend to stop only when the terminal state satisfies one of these two criteria, we cannot be certain of ending the search (save in the obvious case where the exhaustive search graph consists of a finite number of accessible states).

Example. We consider the same traveling salesman problem as before, but now the costs are imprecise, that is to say that, for example, the lengths of the journeys connecting the towns are only imperfectly known. The data are given in Figure 5.2a.

The function $g(n)$ is obtained simply by summing the upper and lower bounds, respectively, of the imprecise costs $[\underline{C}_{ij}, \bar{C}_{ij}]$ along the path corresponding to state n. The function $h(n)$ is a sum of intervals precalculated for each vertex of the graph of towns $(\mathscr{S}, \mathscr{A})$. For vertex i we have

$$[\underline{h}_i, \bar{h}_i] = \widetilde{\min}_{(i,j)\in\mathscr{A}} [\underline{C}_{ij}, \bar{C}_{ij}] \tag{5.2}$$

and for $n = S_0 S_1 \cdots S_j S_k$,

$$\underline{h}(n) = \sum_{l \notin \{S_1, \ldots, S_j, S_k\}} \underline{h}_l, \qquad \bar{h}(n) = \sum_{l \notin \{S_1, \ldots, S_j, S_k\}} \bar{h}_l \tag{5.3}$$

The intervals $[\underline{h}_i, \bar{h}_i]$ are given in Figure 5.2b. Figure 5.2c gives the tree-structure developed by means of the five criteria C1, C23, C2, C3, C4 for the selection of states. The same conventions as for Figure 5.1c have been adopted. Note that the chosen stopping criterion is C23. The algorithm gives the same Hamiltonian path as before (the "son" of state 9) despite the imprecision. Criterion C23 is certainly satisfied for this terminal state, with cost [8, 16], which is indeed the result of using the operator $\widetilde{\min}$ on the ten candidates for development. Note that if we had been content with criterion C2 for stopping the search we would have developed only nine vertices. In the case of the traveling salesman problem, selection by criterion C2 (or C3) clearly comes down to pursuing the algorithm merely with the coefficients $\underline{C}_{ij}$ (or $\bar{C}_{ij}$), that is to say that we are led back to the case of precise data. In particular, if the

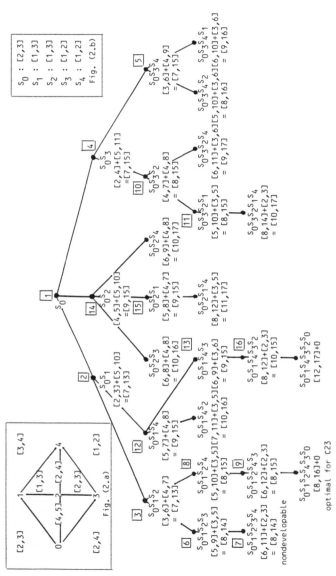

Figure 5.2. Example of search graph with imprecise estimates.

optimal solutions obtained with each of these criteria correspond to the same Hamiltonian circuit (at least), then this Hamiltonian circuit is optimal in the sense of criterion C23, and it is given by the algorithm A^* extended to imprecise data, provided we adopt C23 as a stopping rule. This is what happens in the example. Note that in this example, there is no solution optimal in the sense of C1 (the "son" states of 16 and 9 are terminal states which are not comparable in the sense of C1).

5.1.4. Heuristic Search with Fuzzy Values

Finally let us suppose that the costs making up the evaluation function f are represented by fuzzy intervals as introduced in Chapter 2. This extension of heuristic search methods was suggested by Farreny and Prade [7]. At a given stage of development of the search graph, we have p candidate states $n_1, \ldots, n_p$, where each n_i is associated with a fuzzy interval $\tilde{f}(n_i)$ which constrains the possible values of the function $f(n_i)$. In [7] it is proposed that, to select the state to be developed, n_i should be chosen so that

$$\tilde{f}(n_i) = \widetilde{\min_{j=1,p}} \tilde{f}(n_j) \tag{5.4}$$

Clearly such a state may not exist (cf. Chapter 2, Section 2.2.3.3), in which case any state may be chosen. Obviously (5.4) is a relatively strong condition since it amounts to applying criterion C23 to all the α-cuts of the $\tilde{f}(n_j)$, $j = 1, \ldots, p$.

A different approach involves quantifying the extent to which one evaluation is smaller than another, according to the criteria C1–C4. The $\widetilde{\min}$ operator offers no help in carrying out such quantification, which can, however, be naturally done by using the four indices of comparison of fuzzy intervals introduced in Chapter 3, Section 3.2. These four indices reflect the four criteria C1–C4 used to compare nonfuzzy imprecise evaluations in the preceding paragraph. In the case of ordinary intervals, each of these criteria is satisfied or not; in the case of fuzzy evaluations, the criteria are more or less satisfied.

C1 is to be evaluated by $\mathrm{Nec}(\underline{f}(n_2) > \bar{f}(n_1))$

C2 is to be evaluated by $\mathrm{Nec}(\underline{f}(n_2) \geq \underline{f}(n_1))$

C3 is to be evaluated by $\mathrm{Pos}(\bar{f}(n_2) > \bar{f}(n_1))$

C4 is to be evaluated by $\mathrm{Pos}(\bar{f}(n_2) \geq \underline{f}(n_1))$

where we use the notation of Chapter 3, Section 3.2.1. The criteria C1 and C3 have been made more stringent ($>$ instead of $\geqslant$) for the sake of consistency with that section. Observe that if the membership functions are continuous, then these changes make no difference to the result.

In the case of p candidates for development, the four indices $\mathrm{NS}(\tilde{f}(n_i))$, $\mathrm{NSE}(\tilde{f}(n_i))$, $\mathrm{PS}(\tilde{f}(n_i))$, $\mathrm{PSE}(\tilde{f}(n_i))$ are calculated for each state n_i (see Chapter 3, Section 3.2.3); these express the extent to which $\tilde{f}(n_i)$ is smaller than the remaining evaluations in the sense of C1, C2, C3, and C4, respectively. Next, the states maximizing $\mathrm{NS}(\tilde{f}(n_i))$ are selected. If there are several, the ones maximizing $\mathrm{NSE}(\tilde{f}(n_i))$ are searched, and so on in the above order of the indices, since it can be verified (cf. Section 3.2.2 of Chapter 3) that

$$\mathrm{NS}(\tilde{f}(n_i)) \leqslant \mathrm{NSE}(\tilde{f}(n_i)) \leqslant \mathrm{PSE}(\tilde{f}(n_i))$$

$$\mathrm{NS}(\tilde{f}(n_i)) \leqslant \mathrm{PS}(\tilde{f}(n_i)) \leqslant \mathrm{PSE}(\tilde{f}(n_i))$$

This selection procedure is implemented in the BASIC subroutine given in Section A.1 of the Appendix. Note that the priorities of NSE and PS may be reversed. When this procedure is applied merely to classical intervals, it reduces to the procedure described in the previous paragraph, without criterion C23. As before, the stopping rule for the procedure may be chosen in one of two ways:

- Stop as soon as a terminal state has been chosen.
- Stop only when a terminal state has been chosen for which a condition of the form $E(\tilde{f}(n_i)) \geqslant \Theta$ is satisfied, where E is one of the indices $\{\mathrm{NS}, \mathrm{NSE}, \mathrm{PS}, \mathrm{PSE}\}$ and Θ is a fixed threshold level.

N.B. A different generalization of tree-searching methods results if formula (5.1) for constructing the evaluation function is changed. The properties of the A and A^* algorithms depend, among other things, on the assumption that evaluation functions are additive. However, not all the properties of addition are used, but only associativity, monotonicity, and existence of a zero element. Thus the properties of the algorithms are retained if the evaluation function is changed while preserving the requisite algebraic structure. This will, for instance, be the case if (1) is replaced by $f(n) = \min(g(n), h(n))$ or, more generally, if addition is replaced by a triangular norm or conorm (cf. Chapter 3). See Yager [20] for a study of such generalized A^* algorithms. Their advantage is shown most directly when the evaluation function measures a degree of uncertainty relative to the accessibility of a state on the search graph; the rules for combining such degrees of uncertainty are not necessarily additive. See also Pearl [15].   □

5.2. An Example of Fuzzy Programming: Tracing the Execution of an Itinerary Specified in Imprecise Terms

The procedures and data used by the human brain are not always precisely given. This state of affairs is not a property of the natural language (e.g., English) used to describe procedures or data: both can generally be expressed quite precisely in natural language. Nor is it basically due to the brain's inability to exploit the capabilities of natural language. The source of the imprecision must be sought elsewhere. Sometimes, precise data cannot be obtained. For instance, the time required to perform a process that has not yet been carried out must be estimated in terms of the expected circumstances in which it will be performed. In other cases, high precision is meaningless; this is often the case in domains concerned with human qualities or behavior. For instance, the job description for an advertised position is often specified in very imprecise terms. In many cases precise values used in vague specifications are necessarily more or less arbitrary; vagueness and imprecision seem to be inherent in certain kinds of information used by human beings. Therefore it seems more natural and effective to accept imprecise data as they are. Zadeh [22] has rightly stressed that human reasoning normally uses fuzzy categories:

> Thus, the ability to manipulate fuzzy sets and the consequent summarizing ability constitute one of the most important assets of the human mind as well as a fundamental characteristic that distinguishes human intelligence from the type of machine intelligence that is embodied in present-day digital computers.

Imprecisely specified procedures (called fuzzy algorithms, Zadeh [21, 22]), while more synthetic, have the merit of exhibiting greater adaptability to situations that may be subject to slight perturbations. Other kinds of fuzzy instructions may be imagined (Prade [24]).

An instruction may be fuzzy because it involves fuzzy arguments (if for instance it specifies an operation on fuzzy numbers) or fuzzy functions (e.g., SLIGHTLY-INCREASE, which takes a quantity x into the fuzzy quantity $x \oplus M$, where M is a fuzzy number whose possible values are "small") or, again, fuzzy predicates (i.e., predicates that may have truth values other than "true" or "false") if the instruction is of conditional branching type. A simple example of a program containing such instructions is one to calculate the general ability of a pupil by calculating the average of his marks in science and in literature (each being represented by a possibility distribution on a scale of marks), where it is decided to *slightly increase* the mark for science (for example) if the pupil is *young*. The problems arising in implementing such conditional instructions are briefly touched on in Section 5.2.3.3. Such a program gives a result, in this case the "general ability," in the form of a possibility distribution.

An instruction may also be fuzzy because its arguments, while necessarily taking precise values, are vaguely designated; in the program just described, however, the fuzzy quantities were designated in a nonfuzzy manner. Consider now the two examples:

COMPLETELY-ENUMERATE (HEAVY OBJECTS)
ADD (WEIGHT SMALL BOX) (WEIGHT LARGE BOX)

In the former the program carries out a search in a database of all known more-or-less heavy objects. In the latter, the program seeks a BOX for which the qualifier LARGE is sufficiently plausible, and then a BOX for which the qualification SMALL is equally plausible, and finally adds their weights (themselves perhaps fuzzy); when there is definite ambiguity in designation of the boxes, the program can detect it and stop, returning a message. In both cases it will be necessary to have a filter that evaluates the compatibility between a possible interpretation and the given designation.

In some applications, e.g., robotics, fuzzy instructions can only be carried out if they are given nonfuzzy interpretations. The transition from fuzzy to nonfuzzy is required in order to deal with the real world. This is the case for instructions like

MOVE *A BIT* TO THE LEFT
INCREASE THE PRESSURE BY X

(X is a precise identifier for a fuzzy quantity evaluated elsewhere.)

FIND THE *BIG GREENISH* CUBE

To move from a fuzzy instruction to an executable nonfuzzy instruction one proceeds as follows: for each possibility distribution resulting from the evaluation of a fuzzy element in an instruction, only those actual instances will be used that have sufficient possibility, and then, among these, one interpretation will be adopted (which may later be called into question).

A good example of a sequence of fuzzy instructions that must have nonfuzzy execution is provided by commands intended to guide the movement of a person (or a robot) toward a target. The following itinerary is fairly typical of plans of action used, and communicated, by human beings:

"Continue till you reach a junction about 100 yards away."
"Turn right."
"Continue for about 50 yards till you reach an Asiatic restaurant."
"Turn left."
"Continue for 20–30 yards till you reach a post box."

Such directions bear some resemblance to a computer program. However, a computer program only uses precise operations acting on precise data. Moreover, the interpretation of each instruction of a sequence is unique, and the execution of a sequence is irrevocable. On the other hand, for a sequence of imprecisely specified instructions, those already encountered can be reconsidered (backtracking) if later instructions prove impossible to execute, and will then be interpreted differently (though in accordance with specification).

The problems arising in such "fuzzy programs" are discussed below and illustrated by an example. These problems have been studied by various authors, especially Chang [6], Goguen [9], Tanaka and Mizumoto [16], Uragami *et al.* [17], Imaoka *et al.* [11], Yager [19], Hogle and Bonissone [26], and Farreny and Prade [27].

5.2.1. Execution and Chaining of Instructions

An imprecisely specified instruction can be interpreted in several ways. In some contexts these interpretations can be more or less possible, and so make up a fuzzy set of interpretations. For example, in "Continue till you reach a junction about 100 yards away," the specification "about 100 yards" leads to a fuzzy set of interpretations relative to a given environment. The fuzzy set of distances compatible with "about 100 yards" induces a fuzzy set of "junctions"; in general we shall have a fuzzy set of interpretations.

In order to carry out an instruction, one of its interpretations must be chosen. For simplicity, we here consider only straight sequences of instructions, i.e., those that have no branches (conditional or otherwise). When no interpretation can be found for an instruction of level n, it is necessary to call in question the interpretation chosen for an instruction of level $n - 1$; if a new interpretation can be found for the instruction of level $n - 1$, we can try the instruction of level n again; otherwise we call in question the interpretation chosen for the instruction of level $n - 2$, and so on. Thus either we can follow through the entire sequence of instructions (it is globally interpretable), or else we exhaust the set of possible interpretations for each instruction (in which case the sequence is not globally interpretable).

5.2.1.1. Determining a Set of Interpretations for an Instruction

Each instruction, according to type, may have different kinds of imprecise operand. In the above example, we see operands referring to distances and to directions. The instructions refer to elements to be identified in the environment; thus "reach an asiatic restaurant about 50 yards away" evokes a search for "localities" whose distance is sufficiently compatible with "about 50 yards" and whose kind, at the same time, is sufficiently compatible with "asiatic restaurant." "At the same time" means that the degrees of compatibility with each component of the specification are to be combined: we shall use the min operator for this. In this example, the degree of compatibility measures the possibility that a "locality" corresponds to the specification occurring in the instruction; this will be considered sufficient if it exceeds a given threshold.

Given a specification, the environment will contain a set of interpretations (possibly empty) whose degree of compatibility with the specification exceeds the threshold. It may be that the information about the environment available at the time of executing the instruction represents only a part of this set of interpretations (as happens in the example of Section 5.2.2.2).

5.2.1.2. Choice of Interpretation of an Instruction

In choosing one interpretation among those that are known and sufficiently possible, it is natural to take account of the degree of compatibility attached to each one. Moreover, the nature of the problem may lead to a cost being associated with each possible choice of interpretation. The choice of interpretation actually made will be the result of taking account of both the compatibility and the cost according to a heuristic appropriate to the application.

5.2.1.3. Temporary Failure of an Instruction

When the known set $I(n)$ of sufficiently possible interpretations of an instruction of level n is either empty or has been exhausted [i.e., we cannot carry out the instruction of level $n + 1$ by adopting any of the interpretations in $I(n)$], we have to reconsider the process by which we arrived at $I(n)$:

- By deriving new interpretations of the level-n instruction which were not known before, without calling earlier instructions into question.
- Or by calling into question the results from earlier instructions (backtracking).

5.2.2. Illustrative Example

Figure 5.3 shows the environment in which a robot (initially at $(0, 0)$ and facing in the direction of the arrow) is to execute the itinerary described by the following list:

```
(setq itiner iti1)
= ((reach (target junction) (distance about-100-y)) (turn to-the-right)
(reach (target restaurant asiatic) (distance about-50-y)) (turn
slightly-left) (reach (target post-box on-the-left) (distance
10-to-20-y)))
```

5.2.2.1. General Characteristics

The program given in Section A.2 of the Appendix, of which a run is reproduced in the following section, represents the behavior of a robot

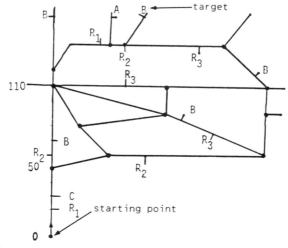

Figure 5.3. Map used in the example. •, Junctions; R_1, Chinese restaurants; R_2, Vietnamese restaurants; R_3, French restaurants; A, driving schools; B, post boxes; C, hairdressers.

according to the following conventions:

- The robot discovers its environment progressively on demand, using two "sensors." One sensor provides interpretations for instructions like "reach": the nature of points of reference (junctions, restaurants, perhaps qualified) and their distances from the robot; the other provides interpretations for instructions like "turn": directions of accessible streets and the distances separating their entrances from the robot.
- The application of these sensors to the environment is simulated: the operator provides the information requested (in the run reproduced below, the operator replies after the prompt "?").
- Each sensor has a limited range, which is set before running the robot program. The robot takes account of these limitations. The robot consults the first sensor only when relevant, i.e., only when it is within a zone of distances compatible with those specified in "reach"-type instructions, taking account of the limited range of the sensors. Moreover, the robot, taking the specification into account, may itself restrict the zone of investigation to less than the full range of the first sensor (i.e., there is no sufficiently compatible interpretation to be found beyond that range).

5.2.2.2. Representation of Imprecise Information

Figure 5.4a–c shows the fuzzy sets corresponding to the specifications of distance, directions and types of restaurant used in the above itinerary. The data provided by the sensors for distance and orientation are supposed to be exact. Any restaurant encountered is supposed to correspond exactly to one of the standard types in a list which is the reference set on which the specification is given. Each fuzzy set corresponding to a continuous space (Figures 5.4a and 5.4b) is represented in the program as a quadruple $(a, b, \Delta a, \Delta b)$.

Underneath Figure 5.4c, showing the fuzzy set of standard types of restaurant [Arabian (A), Chinese (C), French (F), Italian (I), and Vietnamese (V)], is given the representation used in the program in the form of an n-tuple of percentage degrees of possibility for the n alternatives.

5.2.2.3. Sample Run

In the messages, each of the five instructions in the itinerary is referred to as "INSTRUCTION 1 or 2 or $\cdots$ 5."

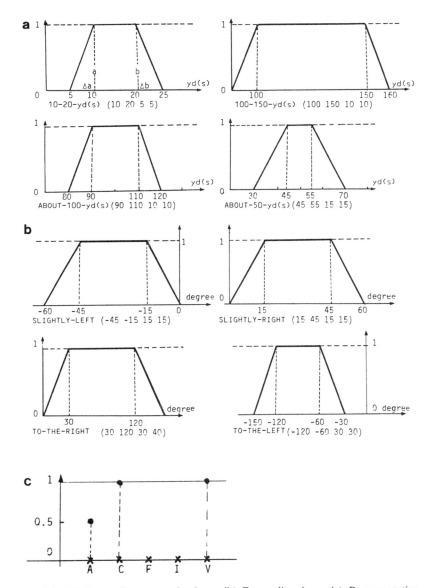

Figure 5.4. (a) Fuzzy distance evaluations. (b) Fuzzy directions. (c) Representation of "Asiatic" (50 100 0 0 100).

Execution of the first itinerary (itil)

1 (exitin)

    * I START WITH INSTRUCTION 1 *
2 I PROCEED FORWARD 85 YARDS
3 I LOOK OUT TO 30 YARDS
4 (junction) ?

5 ((junction (distance 25)))

6 RANKING ((100 (distance 110)))
7 I PROCEED FORWARD 25 YARDS (POSSIB: 100)
    LOCAL SUCCESS INSTRUCTION 1 SINCE
8 I HAVE REACHED junction distance 110 yards

    * I START WITH INSTRUCTION 2 *
9 I LOOK OUT TO 20 YARDS
10 STREETS TO TURN INTO ?

11 ((150 (distance 0)) (130 (distance 0)) (90 (distance 0))
    (-90 (distance 0)) (30 (distance 10)))

12 RANKING ((100 90 (d 0)) (75 130 (d 0)) (100 30 (d 10)))
    I TURN THROUGH 90 DEGREES (POSSIB: 100)
13 LOCAL SUCCESS INSTRUCTION 2 SINCE
    I HAVE TURNED 90 degrees

    * I START WITH INSTRUCTION 3 *
    I PROCEED FORWARD 37 YARDS
14 I LOOK OUT TO 26 YARDS
15 (restaurant asiatic) ?
16

    ((restaurant french (distance 13)))
17

    RANKING ()
18 LOCAL FAIL INSTRUCTION 3
    I RETREAT 37 YARDS

    * I RECONSIDER INSTRUCTION 2*
    I TURN THROUGH 130 DEGREES (POSSIB: 75)
19 LOCAL SUCCESS INSTRUCTION 2 SINCE
    I HAVE TURNED 130 degrees

    * I RECONSIDER INSTRUCTION 3*
    I PROCEED FORWARD 37 YARDS
    I LOOK OUT TO 26 YARDS
    (restaurant asiatic) ?

    no
    RANKING ()
    LOCAL FAIL INSTRUCTION 3
    I RETREAT 37 YARDS

    * I RECONSIDER INSTRUCTION 2*
    I PROCEED FORWARD 10 YARDS
20 I TURN THROUGH 30 DEGREES (POSSIB: 100)
    LOCAL SUCCESS INSTRUCTION 2 SINCE
    I HAVE TURNED 30 degrees

    * I RECONSIDER INSTRUCTION 3*
    I PROCEED FORWARD 37 YARDS
    I LOOK OUT TO 26 YARDS
    (restaurant asiatic) ?

21 ((restaurant chinese (distance 8)) (restaurant vietnamese (distance 18)))

    RANKING ((100 chinese (distance 45)) (100 vietnamese (distance 55)))
    I PROCEED FORWARD 8 YARDS (POSSIB: 100)
    LOCAL SUCCESS INSTRUCTION 3 SINCE
    I HAVE REACHED restaurant chinese distance 45 yards

    * I START WITH INSTRUCTION 4 *
    I LOOK OUT TO 20 YARDS
    STREETS TO TURN INTO ?

    ((-90 (distance 10)) (-45 (distance 18)))

```
    RANKING ((100 -45 (distance 18)))
    i PROCEED FORWARD 18 YARDS
    i TURN THROUGH -45 DEGREES (POSSIB:   100)
    LOCAL SUCCESS INSTRUCTION 4 SINCE
    i HAVE TURNED -45 degrees

    * i START WITH INSTRUCTION 5 *
    i PROCEED FORWARD 8 YARDS
    i LOOK OUT TO 14 YARDS
 22 (post-box on-the-left) ?

 23 ((post-box -90 (distance 13)))

    RANKING ((100 -90 (distance 21)))
    i PROCEED FORWARD 13 YARDS (POSSIB:   80)
    LOCAL SUCCESS INSTRUCTION 5 SINCE
    i HAVE REACHED post-box -90 distance 21 yards

    ITINERARY ENTIRELY TRACED
    = t


        Execution of the second itinerary

 24 (setq itiner iti2)
    = ((reach (target junction) (distance 100-to-150-y)) (turn
    slightly-right) (reach (target restaurant asiatic) (distance about-50-y
    )) (turn on-the-left) (reach (target post-box on-the-left) (distance
    10-to-20-y)))
    (exitin)

    * i START WITH INSTRUCTION 1 *
 25 i PROCEED FORWARD 95 YARDS
 26 i LOOK OUT TO 40 YARDS
    (junction) ?

 27 ((junction (distance 15)) (junction (distance 25)))

 28 RANKING ((100 (distance 110)) (100 (distance 120)))
 29 i PROCEED FORWARD 15 YARDS (POSSIB:   100)
    LOCAL SUCCESS INSTRUCTION 1 SINCE
    i HAVE REACHED junction distance 110 yards

    * i START WITH INSTRUCTION 2 *
    i LOOK OUT TO 20 YARDS
    STREETS TO TURN INTO ?

    ((150 (distance 0)) (130 (distance 0)) (90 (distance 0))
     (-90 (distance 0)) (30 (distance 10)))

 30 RANKING ((100 30 (distance 10)))
 31 i PROCEED FORWARD 10 YARDS
    i TURN THROUGH 30 DEGREES (POSSIB:   100)
    LOCAL SUCCESS INSTRUCTION 2 SINCE
    i HAVE TURNED 30 degrees

    * i START WITH INSTRUCTION 3 *
    i PROCEED FORWARD 37 YARDS
    ! LOOK OUT TO 26 YARDS
    (restaurant asiatic) ?

    ((restaurant chinese (distance 8)) (restaurant vietnamese (distance 18)))

    RANKING ((100 chinese (distance 45)) (100 vietnamese (distance 55)))
    i PROCEED FORWARD 8 YARDS (POSSIB:   100)
    LOCAL SUCCESS INSTRUCTION 3 SINCE
    i HAVE REACHED restaurant chinese distance 45 yards

    * i START WITH INSTRUCTION 4 *
    i LOOK OUT TO 20 YARDS
    STREETS TO TURN INTO ?

    ((-90 (distance 10)) (-45 (distance 18)))

    RANKING ((100 -90 (distance 10)) (50 -45 (distance 18)))
```

```
32  I PROCEED FORWARD 10 YARDS
    I TURN THROUGH -90 DEGREES (POSSIB:   100)
    LOCAL SUCCESS INSTRUCTION 4 SINCE
    I HAVE TURNED -90 degrees

    *  I START WITH INSTRUCTION 5 *
    I PROCEED FORWARD 8 YARDS
    I LOOK OUT TO 14 YARDS
    (post-box on-the-left) ?

    no
    RANKING ()
    LOCAL FAIL INSTRUCTION 5
    I RETREAT 8 YARDS

    *  I RECONSIDER INSTRUCTION 4*
    I PROCEED FORWARD 8 YARDS
33  I TURN THROUGH -45 DEGREES (POSSIB:   50)
    LOCAL SUCCESS INSTRUCTION 4 SINCE
    I HAVE TURNED -45 degrees

    *  I RECONSIDER INSTRUCTION 5*
    I PROCEED FORWARD 8 YARDS
    I LOOK OUT TO 14 YARDS
    (post-box on-the-left) ?

    ((post-box -90 (distance 13)))

    RANKING ((100 -90 (distance 21)))
    I PROCEED FORWARD 13 YARDS (POSSIB:   80)
    LOCAL SUCCESS INSTRUCTION 5 SINCE
    I HAVE REACHED post-box -90 distance 21 yards

    ITINERARY ENTIRELY TRACED
    = t
```

Comments:

1. Start-up of program EXECUTE ITINERARY.

2. Until a displacement of 85 yards has been attained there is no point in "looking" for a "JUNCTION TARGET" since there cannot be sufficient possibility for compatibility between distance traveled and the specification "DISTANCE ABOUT-100-YARDS" (i.e., possibility in excess of the threshold "THRESHOLD" in the program, set at 40%).

3. The range of the sensor associated with "REACH . . . " instructions is set at 40 in this run (see "PCH" in the program). The robot only looks out to 30 yards because beyond this range the total distance traveled (beyond 85 + 30) would no longer be compatible with "DISTANCE ABOUT-100-YARDS" (see the definition of "ABOUT-100-YARDS," and the value of "THRESHOLD").

4, 5. To simulate the sensor for "REACH . . . " instructions the robot inquires whether there is a "JUNCTION" (up to 30 yards away—see 3) and the operator replies with a list (which here has one element) of junctions located. Each element of the list is of the form (JUNCTION (DISTANCE ⟨value-in-yards⟩)).

6. The robot orders the junctions located (only one here) first by nearest distance and then by degree of compatibility with the specifica-

tion (here "ABOUT-100-YARDS"). For each junction it gives the degree of compatibility and then the distance.

7. The robot moves toward the first junction in the ordering (here only one possibility). Following "POSSIB" is the combination (by "min") of the degrees of compatibility between each element of the specification (here a target of type "JUNCTION" and a "DISTANCE"), and the corresponding elements of the action carried out.

8. The robot has traveled 85 + 25 = 110 yards. See the diagram for the position reached.

9. The depth of field of the sensor associated with "TURN . . . " instructions, which locates the nearby streets toward which the robot can turn, is set at 30 yards (see "SEE" in the program).

10, 11. To simulate the sensor associated with "TURN . . . " instructions, the robot inquires whether there are streets to turn into (less than 20 yards ahead, according to 9). The operator replies with a list of which each element has the form (⟨turn-angle-in-degrees⟩ (DISTANCE ⟨value-in-yards⟩)). Here, for example, there are streets at +150°, +130°, +90°, +30° (to the right) and a street at −90° (to the left); the one at +30° is 10 yards ahead.

12. The robot orders the accessible streets (here 5) first by distance and then by degree of compatibility with the instruction (comparison of the turning-angles with the specification "TO-THE-RIGHT," with elimination if the result is less than "THRESHOLD," here 40). For each street it gives the degree of compatibility, the angle of turn, and finally the distance.

13. The robot turns into the first street in the ordering. Following "POSSIB" we find the degree of compatibility between specified and actual directions.

14, 15. The same kind of reasoning as in 2 and 3, respectively.

16, 17. Similar to 4, 5. The operator gives, as in 5, the type of target (here "RESTAURANT") followed by one or more attributes (here just one: "FRENCH") corresponding to those in the request, followed by distance data.

18. Similar to 6, but the one "RESTAURANT" located is the wrong kind.

19. To take the second street in the order of 11, it is sufficient (after returning to the junction) to turn through +40°. Following "POSSIB" we find degree 75 on evaluating directions.

20. To take the third street in the order of 11, we must (following a

retreat) reenter the first street (at $-130°$ to the street along which we retreated) and go forward 10 yards.

21. Similar to 17. Here there are two restaurants at less than 26 yards.

22, 23. Similar to 16, 17. Type of target: "POST BOX"; attribute "-90" (here we reply only "-90" or "$+90$"; we take no account of the direction from which we see the object).

24. *Second itinerary* (same starting point and environment). Observe that, compared with the first, instructions 1, 2, and 4 have slightly altered specifications.

25, 26, 27, 28, 29. Variants of 2, 3, 5, 6, 7, respectively.

30, 31. This time, owing to the specification "SLIGHTLY-RIGHT" instead of "RIGHT," the streets of the first junction are not retained: the robot advances straight away to the second junction (avoiding backtracking).

32, 33. This time a backtrack (see 32) is done, which was avoided in the first itinerary because the latter was more precisely specified: "TURN SLIGHTLY-LEFT" instead of "TURN LEFT."

5.2.3. Problems Arising in Fuzzy Programming

Sections 5.2.1 and 5.2.2 presented the basic principles of executing and chaining fuzzy instructions and an illustrative example was given. In this section we briefly touch on problems to do with "variable" or fuzzy thresholds, imprecise data about the environment, the calculation of fuzzy arguments to fuzzy instructions which require nonfuzzy execution, and "conditional" fuzzy instructions.

5.2.3.1. "Variable" or Fuzzy Thresholds

In the preceding example, only interpretations compatible with specification to a degree exceeding a previously set threshold were retained (all others being definitively rejected), and later they were ordered by degree of compatibility for the purpose of choosing an interpretation. We could envisage "variable" thresholds. This would involve taking an initial relatively high threshold so as to reduce the number of possibilities to consider (e.g., looking for a "junction at about 100 yards" within a zone very close to 100 yards) and then, if the first interpretation is called in question, to possibly lower the threshold in order to extend the set of possible interpretations.

Use of thresholds has the disadvantage of introducing discontinuity into the degrees of compatibility of interpretations of fuzzy instructions, which is precisely what the notion of fuzzy set is supposed to avoid. To alleviate this disadvantage we may use fuzzy thresholds. A fuzzy threshold $\bar{s}$ can be defined as a fuzzy interval of type LR on $[0, 1]$ of the form $\bar{s} = (s, 1, \varepsilon, 0)_{LR}$ (cf. Chapter 2, Section 2.3.1), which is interpreted as follows:

Let F be the fuzzy set associated with the specification, and ω an interpretation of this specification. The fuzzy threshold transforms μ_F to $\mu_{F'}$ such that $\mu_{F'} = \mu_{\bar{s}} \circ \mu_F$. Then

- If $\mu_F(\omega) \geq s$ then $\mu_{F'} = 1$ (interpretations compatible with F to level at least s are considered equivalent).
- If $\mu_F(\omega) < s - \varepsilon$ then $\mu_{F'}(\omega) = 0$ (interpretations compatible to level less than $s - \varepsilon$ are rejected).
- Intermediate levels of compatibility are more or less modified according to the form of the function L. A threshold of level λ (in the usual sense) is expressed as $\bar{s} = (1, 1, 1 - \lambda, 0)_{LR}$ with $L(u) = 1 - (1 - \lambda)u$ for $0 < \lambda \leq 1$ if $u \leq 1$, and $L(u) = 0$ if $u > 1$. The discontinuity arising with classical thresholds can be eliminated by choosing L to be continuous and such that

$$\lim_{u \to 1} L(u) = 0$$

5.2.3.2. Imprecise Observation of the Environment

In Sections 5.2.1 and 5.2.2 we have implicitly supposed that the human or the robot has correct and precise information about the environment, i.e., (in the example) having exact knowledge of distances and angles in various directions, the restaurants also belonging definitely to standard categories. Only the specifications in the instructions were liable to be imprecise.

Now suppose that the distances and directions are evaluated in imprecise or fuzzy fashion, as represented by possibility distributions. A restaurant also may not clearly correspond to a single standard type, but belong to some degree to several; then the perception of a restaurant can be represented as a possibility distribution over the standard types. For example, one restaurant that could be described as "European" and another perceived as "vaguely Chinese" could be represented by possibility distributions such as those in Figure 5.5.

The degree of compatibility of an imprecise evaluation with an

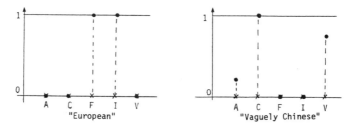

Figure 5.5. Representation of "European" and "Vaguely Chinese."

imprecise specification (represented by the possibility distribution μ_E and the fuzzy set S, respectively, defined on the same reference set U) can be estimated by two scalar measures:

- The degree of possibility that the evaluation corresponds to the specification, defined by

$$\Pi(S; E) = \sup_{u \in U} \min(\mu_S(u), \mu_E(u)) \qquad (5.5)$$

- The degree of necessity (or certainty, if preferred) that the evaluation corresponds to the specification, defined by

$$N(S; E) = \inf_{u \in U} \max(\mu_S(u), 1 - \mu_E(u)) \qquad (5.6)$$

These two quantities correspond, respectively, to the possibility and necessity of a fuzzy event (cf. Chapter 1, Section 1.7). $\Pi(S; E)$ evaluates the extent to which the possibility distributions μ_S and μ_E have a part in common, and $N(S; E)$ the extent to which μ_S completely "covers" μ_E. We have that $N(S; E) \leq \Pi(S; E)$. When the evaluation E is precise, i.e., corresponds to a singleton $\{u_0\}$ of U, we of course have that $N(S; E) = \Pi(S; E) = \mu_S(u_0)$, which was the case in Sections 5.2.1 and 5.2.2.

The possibility measure $\Pi(S; E)$ and the necessity measure $N(S; E)$ play an important role in searching a database containing imprecise or fuzzy information by means of a filter that represents a query with possibly imprecise or fuzzy specification (see also Cayrol, Farreny, and Prade [3, 4] and Chapter 6).

In imprecise perception of the environment we have two numbers $\Pi(S; E)$ and $N(S; E)$ to estimate compatibility between a possible interpretation and a specification, instead of just one in the case of precise perception. Thus the different possible interpretations can be ordered according to their degrees of possibility, the degrees of certainty serving to break ties between interpretations of equal possibility. In particular, when the specification is imprecise but not fuzzy (i.e., when S corresponds to a nonfuzzy subset of U), $N(S; E)$ can be nonzero only if $\Pi(S, E)$ is unity. Note also that the degree of necessity evaluates the certainty that a possible interpretation satisfies the specification, and not the certainty that this interpretation is "correct"; i.e., it has to do with whether the succeeding instructions can be successfully executed. In fact there may be several distinct interpretations perfectly compatible with a given specification.

The procedure for execution and sequencing of instructions that was set out in Section 5.2.1 and illustrated in 5.2.2 can be extended to the case of imprecise perception of the environment, using the possibility measure $\Pi(S; E)$ as principal criterion for ordering interpretations, the necessity measure $N(S; E)$ being used as a secondary criterion for refining the ordering. In the example, if the distances are imprecisely perceived, one may be led to consider a greater number of junctions which might be thought (given the available information) to be at the specified distance. In general, the fact of having only an imprecise perception of the environment can only increase the number of possible interpretations of an instruction.

The program given in the Appendix calculates degrees of possibility when, in the example, there is imprecise perception of the nature of the restaurants (evaluation of the possibility of a fuzzy event on a discrete reference set).

5.2.3.3. Imprecise Specifications Resulting from Calculation

In the example of Section 5.2.2, the imprecise specifications which were represented by possibility distributions were given explicitly in the instructions. In general, such specifications may be obtained by calculations which themselves involve instructions with fuzzy operands or fuzzy functions and which yield possibility distributions as their results. The existence of methods of calculation for fuzzy numbers (cf. Chapter 2) means that there is no particular problem in implementing by computer the execution of instructions involving computations on imprecise specifications, except perhaps for conditional branching. We now briefly consider this question.

Consider for example the following test:

$$\text{If } X \geq Y \text{ then routine 1}$$
$$\text{else routine 2}$$

where X and Y may be two fuzzy quantities. The possibility $\text{Pos}(\bar{X} \geqslant Y)$ that X is at least as great as Y can be calculated, as also the possibility $\text{Pos}(\bar{Y} > X)$ of the contrary event. We have

$$\text{Pos}(\bar{X} \geqslant Y) = \sup_{\substack{u,v \\ u \geqslant v}} \min(\pi_X(u), \pi_Y(v)) \tag{5.7}$$

and

$$\text{Pos}(\bar{Y} > X) = \sup_{\substack{u,v \\ u < v}} \min(\pi_X(u), \pi_Y(v)) \tag{5.8}$$

where $\pi_X = \mu_F$ and $\pi_Y = \mu_G$ are the possibility distributions for the fuzzy sets F and G of the more or less possible values of X and Y (cf. Chapter 3, Section 3.2.2). Note that $\text{Pos}(\bar{X} \geqslant Y) = 1$ or $\text{Pos}(\bar{Y} > X) = 1$ since one at least of two contradictory alternatives must be completely possible. Suppose that routine 1 consists of calculating Z with result represented by the possibility distribution $\pi_Z^{(1)}$, and routine 2 of another way of calculating Z with result the possibility distribution $\pi_Z^{(2)}$. The possibility distribution for the more or less possible values of Z will then be given, in the case of routine 1, by

$$\pi_Z(\omega) = \min(\text{Pos}(\bar{X} \geqslant Y), \pi_Z^{(1)}(\omega)) \tag{5.9}$$

and for routine 2 by

$$\pi_Z(\omega) = \min(\text{Pos}(\bar{Y} > X), \pi_Z^{(2)}(\omega)) \tag{5.10}$$

Further, if routines 1 or 2 use values of X or Y, the possibility distributions occurring in the calculation of $\pi_Z^{(1)}$ or $\pi_Z^{(2)}$ should be consistent with the constraint $X \geqslant Y$ (for routine 1) or $X < Y$ (for routine 2). For routine 1, we use

$$\begin{aligned} \pi_X^{\text{modified}} &= \mu_{F \cap [G, +\infty)} \\ \pi_Y^{\text{modified}} &= \mu_{G \cap (-\infty, F]} \end{aligned} \tag{5.11}$$

and in the case of routine 2

$$\begin{aligned} \pi_X^{\text{modified}} &= \mu_{F \cap (-\infty, G[} \\ \pi_Y^{\text{modified}} &= \mu_{G \cap]F, +\infty)} \end{aligned} \tag{5.12}$$

where the fuzzy intervals $[G, +\infty)$, $(-\infty, F]$, $(-\infty, G[$ and $]F, +\infty)$ are defined as in Section 3.2.1 of Chapter 3, and where the intersection is defined by min. If it is later necessary to recombine the results of routines 1 and 2 for Z, this can be done by the fuzzy set-theoretic union (using max) of the results.

For a more detailed discussion of the problems arising with fuzzy conditional instructions, the reader may consult Chang [5] and Adamo [1]. We now close this section concerning various questions arising in fuzzy programming by a short discussion of fuzzy conditional branching in a program which, instead of calculating possibility distributions, imprecisely specifies actions to be executed.

5.2.3.4. Conditional Instructions in an Imprecisely Specified Procedure

The itinerary of Section 5.2.2 was given as a series of instructions in which there was no instance of a conditional branching instruction. This kind of imprecisely specified procedure may in general give rise to tests of the form

$$If \ \langle condition \rangle \ then \ \langle action \ 1 \rangle$$
$$else \ \langle action \ 2 \rangle$$

The condition and the actions 1 and 2 may have imprecise or fuzzy specifications. Then action 1 should be associated with the degree of possibility that the condition is satisfied, and action 2 with the degree of possibility that the condition is not satisfied. If the condition is of the form "X is S," where μ_S is a possibility distribution on U, and the current evaluation of X is represented by μ_E, these two degrees of possibility are represented, respectively, by $\Pi(S; E)$ and $1 - N(S; E)$ [cf. formulas (5.5) and (5.6)]. Then the action corresponding to the greater degree of possibility is chosen. If the execution of this instruction or of subsequent ones should fail (the various possible interpretations having been explored) then one may be led to backtrack to an action not retained in the first instance (provided its degree of possibility is adequate).

5.2.4. Concluding Remarks

The execution of an imprecisely specified procedure is, as we have seen, guided by estimation of the compatibility between the current situation as perceived, and the specification. This estimation of compatibility plays an important role in many other problems of artificial intelligence where imprecision and uncertainty are involved, as well as in execution of programs:

1. Search for applicable rules in a given situation in an expert system (where the rules may be imprecisely specified and the situation

imprecisely known), and calculation of possibility and necessity
that conditions occurring in a rule "if . . . then . . . " are satisfied
(cf. Chapter 4).
2. Searching a database for items possibly or necessarily satisfying
   conditions expressed in a query (cf. Chapter 6).
3. Pattern recognition of objects imprecisely described (cf. Farreny
   and Prade [8], Dubois and Jaulent [25]).

It should be observed that questions different from those considered
here arise in connection with itinerary problems in artificial intelligence.
As well as the execution of a series of instructions in a real environment,
we also have problems such as finding an imprecisely specified route on a
map (also leading to problems of compatibility estimation) or generating
a series of directions for arriving at a destination, given an imprecise map
(see McDermott and Davis [12]).

Appendix: Computer Programs

A.1. Selection of "the Smallest" of *N* Fuzzy Numbers

```
REM ==================================================================
REM =                       SUBROUTINE MIN'N'NBF                      =
REM =                                                                 =
REM = Parameters : NBRCOMP,INCOMP%( ),PTRCOMP%( ),INDSEL,PTRSEL       =
REM = The subroutine selects the smallest among NBRCOMP fuzzy numbers.=
REM = These numbers must be indexed, their indices range in INDCOMP%( )=
REM = and the pointers for accessing these numbers are in PTRCOMP%( ).=
REM = Output : INDSEL index of the selected number                    =
REM =          PTRSEL pointer for accessing the selected number       =
REM =                                                                 =
REM ==================================================================
MIN'N'NBF:
        PRINT "NBRCOMP= ",NBRCOMP
REM If there is only one number, it is selected :
REM ----------------------------------------
        IF NBRCOMP=1 THEN INDSEL=INDCOMP%(1) : &
                         PTRSEL=PTRCOMP%(1) : &
                         RETURN
REM Otherwise each number is compared with the extended minimum of the others :
REM -------------------------------------------------------------------
        FOR I=1 TO NBRCOMP
          X=MAXFLOU
          Y=R
          GOSUB AFFECT'NBF
          Z=R
          FOR J=1 TO NBRCOMP                              ! consultation min Aj
            IF I<>J THEN X=PTRCOMP%(J) : GOSUB MIN'NBF !         j<>i
          NEXT J
          X=PTRCOMP%(I)
          GOSUB COMP'NBF
          TABCOMP(I,1)=1-PSE
          TABCOMP(I,2)=1-PS
          TABCOMP(I,3)=1-NSE
          TABCOMP(I,4)=1-NS
        NEXT I
```

```
REM  Selection of the smallest number :
REM  -------------------------------
        EPS=0.0001                                    ! tolerance for the
        FOR IND=1 TO 4                                ! selection
          GOSUB SELECTION
          IF NB'SELEC=1 THEN IND=4
        NEXT IND
        FOR I=1 TO NBRCOMP
          ! selectionner alors le premier parmis ceux qui restent
          IF INDCOMP%(I)<>0 THEN INDSEL=INDCOMP%(I) : &
                                  PTRSEL=PTRCOMP%(I) : &
                                  I=NBRCOMP
        NEXT I
        RETURN


REM  Subroutine for selecting the smallest number(s) with respect to one of
REM  the indices IND. The indices of the numbers which are not selected re-
REM  ceive the value zero.

SELECTION:
    VAL'IND=0
    ! The maximum value taken by the index considered is determined :
    FOR I=1 TO NBRCOMP
      IF INDCOMP%(I)<>0 THEN VAL'IND=VAL'IND MAX TABCOMP(I,IND)
    NEXT I
    NB'SELEC=0
    FOR I=1 TO NBRCOMP
      ! All the numbers having the same value for the index  considered
      ! (up to the tolerance EPS) are selected.

      IF INDCOMP%(I)<>0 THEN &
        IF TABCOMP(I,IND)<=VAL'IND-EPS THEN INDCOMP%(I)=0 &
                                  ELSE NB'SELEC=NB'SELEC+1
    NEXT I
    PRINT #1
RETURN
```

A.2. Tracing Imprecisely Specified Itineraries

```
;The following program corresponds to the working session presented in 5.2.2.3.
;It is written and tested in LISP on Macintosh (512 K, dialect LELISP). To run
;the program: call (EXITIN), after having loaded the itinerary on the atom
;ITINER. For instance, do: (SETQ ITINER ITI1) or (SETQ ITINER ITI2) to obtain one
;of the sessions presented in the text

;Initializations. To be modified according to your wishes.
;THRESHOLD: for grades of compatibility (see 5.2.1). See FPMLAND,FPMDIR.
;DEPTHSHOT: To limit perception ("vision") when executing instructions of the
;"REACH" type. See POSDIST.
;NEIGH: neighbourhood. To limit perception ("vision") when executing instructions
;of the "TURN" type. See EYEDIR.
;Fuzzy sets defined on continuous universes are represented by 4-tuples
;(see 5.2.2.2) prefixed by T. Those defined on discrete universes are represented
;in extension, but prefixed by ().
(SETQ threshold 40 depthshot 40 neigh 20
        to-the-right '(T (30 120 30 40))
        on-the-left '(T (-120 -60 30 30))
        slightly-right '(T (15 45 15 15))
        slightly-left '(T (-45 -15 15 15))
        chinese '(() (0 100 0 0 0))
        vietnamese '(() (0 0 0 0 100))
        fuzchinese '(() (25 100 0 0 75))
        french '(() (0 0 100 0 0))
        asiatic '(() (70 100 0 0 100))
        european '(() (0 0 100 100 0))
        about-100-y '(T (90 110 10 10))
        10-to-20-y '(T (10 20 5 5))
        100-to-150-y '(T (100 150 10 10))
        about-50-y '(T (45 55 15 15)))

(DE printj (l)
```

```
;prints one sublist per line
(MAPCAR '(LAMBDA (x) (PRINT x)) l) (TERPRI))


;Two examples of itineraries corresponding to the working session presented in
;2.2.c. Before running EXITIN - call (exitin) - apply SETQ to ITINER.
(SETQ iti1 '((reach (target junction) (distance about-100-y))
                        (turn to-the-right)
                        (reach (target restaurant asiatic) (distance about-50-y))
                        (turn slightly-left)
                        (reach (target post-box on-the-left) (distance 10-to-20-y))))

(SETQ iti2 '((reach (target junction) (distance 100-to-150-y))
                        (turn slightly-right)
                        (reach (target restaurant asiatic) (distance about-50-y))
                        (turn on-the-left)
                        (reach (target post-box on-the-left) (distance 10-to-20-y))))
(DE exitin ()
;EXecution of ITINerary. This is the main procedure - called by (exitin) -.
;NCHAIN: index of current chain segment. Pointer for proceeding into the set of
;instructions pertaining to an itinerary.
;LABELSTACK: stack of labels attached to each instruction.
;The indices of labels are ranked in the reverse order with respect to the
;instructions.
 (PROG (nchain labelstack nstack aux ox)
   (SETQ nchain 1 aux 1)
   (WHILE (<= aux (LENGTH itiner))
                (SETQ labelstack (CONS (SETQ ox (GENSYM)) labelstack) aux (1+ aux))
                (PUTPROP ox 0 'dist)
                (PUTPROP ox 0 'turn))
   (WHILE (AND (<= nchain (LENGTH itiner)) aux)
                (TERPRI)
                (PRINT "* I START WITH INSTRUCTION " nchain " *")
                (COND ((turn? nchain) (perdir nchain)))
                (COND ((trycumex nchain) (SETQ nchain (1+ nchain)))
                          (T (SETQ aux ()))))
   (TERPRI)
   (COND (aux (PRINT "ITINERARY ENTIRELY TRACED") T)
             (T (PRINT "I DON'T TRACE THE ITINERARY") ()))))


(DE trycumex (nchain)
;TRY CUMulative EXecution of rank NCHAIN instruction. Called by EXITIN.
;NSTAC: instruction pointer to trace backtrackings.
 (PROG (nstac)
   (SETQ nstac nchain)
   (WHILE (AND (<= 1 nstac) (<= nstac nchain))
                (COND ((tryindex nstac)
                           (succinst nstac)
                           (SETQ nstac (1+ nstac))
                           (COND ((<= nstac nchain)
                                     (recons nstac)
                                     (cond ((turn? nstac) (perdir nstac))))))
                          (T (failinst nstac)
                              (SETQ nstac (1- nstac))
                              (recons nstac))))
   (< nchain nstac)))


(DE succinst (rinst)
;In case of SUCCessful execution of ranked RINST INSTruction. Called by TRYCUMEX.
 (PRINT "LOCAL SUCCESS INSTRUCTION " rinst " SINCE")
 (messinst rinst))


(DE failinst (rinst)
;In case of FAILure of ranked RINST INSTruction. Called by TRYCUMEX.
 (PRINT "LOCAL FAIL INSTRUCTION " rinst)
 (condretreat rinst))


(DE nlabel ()
;Returns the label of rank RINST (global variable) found in the stack LABELSTACK.
;Called by TIETURN, TIEFORWARD, PERDIR, etc ...
 (NTH (1- rinst) labelstack))


(DE recons (rinst)
;Called by TRYCUMEX
 (TERPRI)
 (PRINT "* I RECONSIDER INSTRUCTION " rinst '*))
```

```
(DE tryindex (rinst)
;TRY INDependent EXecution of the ranked RINST instruction. Called by TRYCUMEX.
 (COND ((turn? rinst) (tieturn rinst))
        (T (tieforward rinst))))


(DE tieturn (rinst)
;Tries Independent Execution of the rank RINST type "TURN" instruction. Called
;by TRYCUMEX.
;Property MATCHDIR: stores the MATCHing DIRections (i.e. compatible with the
;current instruction).
;NB  EXITIN and TRYCUMEX order a perception act (PERDIR) before each call to
;TIETURN.
;Property ACT: stores the trace of the turn act (subsequently used by MESSINST)
;Property DIST: updating according to the walked DISTance for the current
;instruction (initialized in EXITIN).
 (PROG (l m n)
  (COND ((NULL (SETQ l (GETPROP (nlabel) 'matchdir))) ())
        (T (SETQ m (- (distnextarget l) (distalready rinst))))
;M contains the difference between the ALREADY walked DISTance for the current
;instruction and the DISTance to the NEXT TARGET (accessible street). If M is
;nonzero one must first delete the preceding angle, then proceed by M steps, and
;lastly turn according to the value prescribed in the instruction.
;If M is 0, only the angle value needs be corrected.
;NB: there is no use stepping backwards (when trying another street within a
;single instruction) since the adopted heuristic requires trying the closest
;compatible streets first (ties are broken according to compatibility degrees)
                (COND ((<> 0 m)
                        (COND ((<> 0 (SETQ n (turnalready rinst)))
                                (PRINT "! TURN THROUGH " (- 0 n) " DEGREES")))
                        (PRINT "! PROCEED FORWARD " m " YARDS")
                        (PUTPROP (nlabel) (CADR (RAC (CAR l ))) 'dist)
                        (SETQ n (CADAR l ))
                        (T (SETQ n (- (CADAR l) (turnalready rinst))))))
                (PRINT "! TURN THROUGH " n " DEGREES " (LIST "POSSIB: " (CAAR l )))
                (PUTPROP (nlabel) (CADAR l) 'v)
                (PUTPROP (nlabel) (LIST (CADAR l) 'degrees) 'act)
                (PUTPROP (nlabel) (CDR l) 'matchdir)
                T))))

(DE turnalready
;To capture the value (degrees) of the preceding turn. Called by TIETURN.
 (rinst) (GETPROP (nlabel) 'turn))


(DE turn?
;To know whether rank RINST instruction is a TURN.
;Called by EXITIN, TRYCUMEX, T ETURN.
 (rinst) (EQUAL (CAR (NTH (1- rinst) itiner)) 'turn))


(DE perdir (rinst)
;PERception of DiRections (of streets) compatible with the interpretation of
;ranked instruction. EYEDIR will evaluate (simulation by calling the operator)
;the accessible directions (OPENDIR), then FPMDIR will retain only the ones
;compatible with the instruction. They are ranked first according to the
;distance, and then according to compatibility degrees.
;Called by TRYCUMEX and EXITIN.
 (PROG (opendir x)
  (SETQ opendir (eyedir))
  (COND ((NOT (NULL opendir))
          (PUTPROP (nlabel) (SETQ x (fpmdir (CADR (NTH (1- rinst) itiner)) opendir))
'matchdir)))
  (PRINT "RANKING " x)))

(DE eyedir ()
;EYE DiRections. Called by PERDIR. To simulate (call to the operator) the
;accessible street sensor (see session in 5.2.2.3 for the format of responses.
 (PROG (i)
  (PRINT "! LOOK OUT TO " neigh " YARDS")
  (PRINT "STREETS TO TURN INTO ?")
  (TERPRI)
  (COND ((NOT (MEMBER (SETQ i (READ)) '(no n any)))
          (TERPRI)
          i))))

(DE perland (rinst !depthshot)
;PERception of LANDmarks: for type "REACH" instruction. Called by TIEFORWARD.
;EYEDIR controls the simulation of perception. TRANSLIST recomputes the
```

```
;perceived distances (relative to the already reached point) taking into account
;the already walked distance.
;FPMLAND retains the compatible landmarks (according to attributes except
;distance) and ranks them (according to the travel heuristic: the nearest first;
;ties are broken following the compatibility degrees with regard to other
;attributes).
 (PROG (openland x)
  (SETQ openland (translist (distalready rinst) (eyeland rinst)))
  (COND ((NOT (NULL openland))
         (PUTPROP (nlabel)
                              (SETQ x (fomland (CDDADR (NTH (1- rinst) itiner)) (mapct 'CDR
openland)))
                              'matchland)))
  (PRINT 'RANKING ' x)))
;TRANSlation of LISTs: see PERLAND.
  (mapct '(LAMBDA (x) (elemtrans base x)) 1))

(DE elemtrans (base n)
;ELEMentary TRANSlation. Called by TRANSLIST.
  (COND ( (NULL (CDR n)) (LIST (LIST 'distance (+ base (CADAR n)))))
        (T (CONS (CAR n) (elemtrans base (CDR n))))))

(DE eyeland (rinst)
;EYE LANDmarks. To simulate the landmark sensor necessary for type "REACH"
;instructions. LDEPTHSHOT is a global variable declared in PERLAND. It is the
;useful distance to landmarks (see the call of PERLAND in TIEFORWARD, see the
;role of POSTDIST). Called by PERLAND.
  (PROG (l)
    (COND ((= 0 ldepthshot) (PRINT "USELESS TO LOOK OUT"))
          (T (PRINT "! LOOK OUT TO " ldepthshot " YARDS")))
    (PRINT (CDADR (NTH (1- rinst) itiner)) " ?")
    (TERPRI)
    (COND ((NOT (MEMBER (SETQ l (READ)) (no n any))) (TERPRI l))))

(DE tieforward (rinst)
;Tries independent Execution of "step FORWARD" instructions (i.e. "REACH" type);
;called by TRYINDEX and TIEFORWARD. When beginning the interpretation of the ranked
;RINST instruction, there is no sensing order yet (type "PERLAND"): the
;second part of TIEFORWARD will determine the minimal move to do before calling
;PERLAND (to reach a distance superior to the THRESHOLD of compatibility with the
;distance specified in the instruction). Moreover, this move will be limited to
;the point located at a distance DEPTHSHOT from the point beyond which there
;is no longer compatibility with the distance specified in the instruction. After
;calling PERLAND, TIEFORWARD makes a recursive call to itself. The first half of
;TIEFORWARD computes the distance to the next landmark, orders the move (always
;forward: the heuristic favors the closest landmarks), and prints the minimum
;between the grade of compatibility located ahead of each landmark descriptor
;selected by FPMLAND and the grade of compatibility of the corresponding
;distances. The procedure FPMLAND only works on degrees of compatibility of
;attributes which are not distances.
  (PROG (l m n x y)
   (COND ((NOT (NULL (SETQ l (GETPROP (nlabel) 'matchland))))
          (SETQ m (- (SETQ n (distnextarget l)) (distalready rinst)))
          (COND ((< 0 m)
                 (PRINT "! PROCEED FORWARD " m " YARDS "
                  (LIST 'POSSIB: "
                    (MIN (CAAR l)
                         (compcont n
                          (CADR (EVAL (CADAR (CDDR (NTH (1- rinst) itiner))))))))))
                 (PUTPROP (nlabel) (prepmsg (CDAR l) n) 'act)
                 (PUTPROP (nlabel) (CDR l) 'matchland)
                 T)
                (T (SETQ m (posdist (SETQ x (CADR (EVAL (CADAR (CDDR (NTH (1- rinst) itiner))))))
                               (GETPROP (nlabel) 'dist)
                               depthshot))
                   (COND ((< 0 m)
                          (PRINT "! PROCEED FORWARD " m " YARDS")
                          (PUTPROP (nlabel) (SETQ y (+ (GETPROP (nlabel) 'dist) m)) 'dist)
                          (perland rinst (posdist x y 0))
                          (tieforward rinst))))))))

(DE distalready
;DISTance ALREADY walked due to the preceding interpretations of rank RINST
;instruction. Called by TIEFORWARD and TIETURN.
  (rinst) (GETPROP (nlabel) 'dist))
```

```
;DE distnextarget
;DISTance to the NEXT TARGET. Called by TIEFORWARD and TIETURN.
  (I) (CADR (rac (CAR I))))
(DE prepmess (I m)
;PREPares MESSages. Called by TIEFORWARD.
  (REVERSE (CONS 'YARDS
                 (CONS m
                       (CONS 'DISTANCE
                             (APPEND (CDR (REVERSE I))
                                     (LIST (CADADR (NTH (1- rinst) itiner)))))))))

(DE posdist (distinst distalready x)
;POSsible DISTance. Called by TIEFORWARD. Compares the fuzzy distance (4-tuple
;denoted DISTINST) deriving from a type "REACH" instruction, with the precise
;distance (DISTALREADY) already walked. Upon calling X receives 0 or DEPTHSHOT.
;0 if called by PERLAND (from TIEFORWARD), DEPTHSHOT if directly called by
;TIEFORWARD. In the latter case, assuming that the distance already walked is
;compatible with the instruction, one refrains from proceeding beyond a distance
;DEPTHSHOT from the upper bound of the distance compatible with the instruction.
  (PROG (delta)
    (COND ((<= (compcont distalready distinst) threshold)
           (SETQ delta (- (CAR distinst) (CADDR distinst)))
           (WHILE (<= (compcont (+ distalready delta) distinst) threshold)
                  (SETQ delta (1+ delta)))
          delta)
          (T (SETQ delta 0)
             (WHILE (AND (<= delta depthshot)
                         (<> (compcont (+ distalready (+ x delta)) distinst) threshold))
                    (SETQ delta (1+ delta)))
             (COND ((<= 0 delta) 0)
                   (T (SETQ delta (1- delta))))))))

(DE messinst (rinst)
;MESSages sent when the execution of an INSTruction is achieved. Called by
;SUCCINST
  (PRIN "I HAVE ")
  (COND ((turn? rinst) (PRIN "TURNED "))
        (T (PRIN "REACHED ")))
  (MAPC '(LAMBDA (x) (PRIN x) (PRIN " "))
        (GETPROP (nlabel) 'act))
  (TERPRI))

(DE condretreat (rinst)
;RETREAT if needed. Called by FAILINST.
  (PROG (I)
    (COND ((<> 0 (SETQ I (GETPROP (nlabel) 'dist)))
           (PRINT "I RETREAT " I " YARDS")
           (PUTPROP (nlabel) 0 'dist)))))

(DE fpmdir (parinst opendir)
;Fuzzy Pattern Matching on DIRections (i.e. angles). Called by PERDIR. PARINST
;is the PART of information deriving from the INSTruction, OPENDIR is the list of
;OPEN DIRections. FPMDIR ranks the matching directions: the nearest first, ties
;being broken by means of the grades of compatibility (between angles).
  (PROG (aux)
    (clas (mapct '(LAMBDA (x)
                    (COND ((<> (SETQ aux (compcont (CAR x) (CADR (EVAL parinst)))) threshold)
                           (CONS aux x)))
                 opendir))))

(DE compcont (x I)
;COMParison between a fuzzy set and a precise value, on CONTinuous universes.
;Called by FPMDIR, POSDIST, COMPLAND. X is the precise element obtained from
;sensors, L is the representation of the fuzzy set deriving from the instruction.
  (COND ((>= x (+ (CADR I) (rac I))) 0)
        ((<> x (CADR I))
         (- 100 (DIV (* 100 (- x (CADR I))) (rac I))))
        ((>= x (CAR I)) 100)
        ((<> x (- (CAR I) (CADDR I)))
         (- 100 (DIV (* 100 (- (CAR I) x)) (CADDR I))))
        (T 0)))

(DE clas (I)
;CLASsifies the elements in list L. Called by FPMDIR, FPMLAND.
  (PROG (x)
  (COND ((NULL I) ())
```

```
                ((NULL (CDR l)) l)
                (T (CONS (SETQ x (choice l)) (clas (remove x l)))))))

(DE remove (x l)
;REMOVEs the first occurrence of element X in list L at the top-level. Called by
;CLAS.
  (COND ((NULL l) ())
        ((EQUAL x (CAR l)) (CDR l))
        ((CONS (CAR l) (remove x (CDR l))))))

(DE choice (l)
;Called by CLAS for a CHOICE in L.
  (PROG (x)
    (COND ((NULL (CDR l)) (CAR l))
          ((< (CADR (rac (CDR l)))
              (CADR (rac (SETQ x (choice (CDR l))))))
           (CAR l))
          ((= (CADR (rac (CAR l))) (CADR (rac x)))
           (COND ((>= (CAAR l) (CAR x)) (CAR l))
                 (T x)))
          (T x))))

(DE fpmland (parinst openland)
;Fuzzy Pattern Matching for LANDmarks. Called by PERLAND. PARINST is the PART of
;information deriving from the INSTruction; OPENLAND is the list of accessible
;landmarks (according to sensors). FPMLAND compares only attributes other than
;distance. The results are ranked according to the distances (nearest first);
;ties are broken by means of the grades of compatibility for other attributes.
  (PROG (aux)
    (clas (mapct '(LAMBDA (x)
                    (COND ((> (SETQ aux (compland (CDR (REVERSE x)) (REVERSE parinst)))
                              threshold)
                           (CONS aux x))))
                 openland))))

(DE compland (land ins)
;COMParison of LANDmarks. Called by FPMLAND. LAND is obtained from sensors and
;INS from the instruction.
  (PROG (x)
    (COND ((NULL ins) 100)
          (T (MIN (COND ((CAR (SETQ x (EVAL (CAR ins))))
                         (compcont (CAR land) (CADR x)))
                        (T (compdis (CADR (EVAL (CAR land))) (CADR x))))
                  (compland (CDR land) (CDR ins)))))))

(DE compdis (per ins)
;COMParison of two fuzzy sets on DIScrete universes. Called by COMPLAND. PER
;is obtained from sensors and INS from the instruction.
  (COND ((NULL per) 0)
        (T (MAX (MIN (CAR per) (CAR ins))
                (compdis (CDR per) (CDR ins))))))

(DE rac (l)
;Returns the last element of list L at the top-level.
  (CAR (REVERSE l)))

(DE mapct (ff ll)
;Applies the FF function to each element of list LL. Returns the list of results
;different from (). Called by FPMDIR, FPMLAND, MAPCT ...
  (PROG (auxx)
    (COND ((NULL ll) ())
          ((SETQ auxx (EVAL (LIST ff '(CAR ll))))
           (CONS auxx (mapct ff (CDR ll))))
          ((mapct ff (CDR ll))))))
```

References

1. ADAMO, J. M. (1980). L.P.L—A fuzzy programming language 1. Syntactic aspects. *Fuzzy Sets Syst.*, **3**(2), 151–179; 2. Semantic apsects, *ibid.*, **3**(3), 261–289.
2. AGUILAR-MARTIN, J., and LOPEZ DE MANTARAS, R. (1982). The process of classification and learning the meaning of linguistic descriptors of concepts. In *Approximate Reasoning in Decision Analysis* (M. M. Gupta and E. Sanchez, eds.), North-Holland, Amsterdam, pp. 165–175.

3. CAYROL, M., FARRENY, H., and PRADE, H. (1980). Possibility and necessity in a pattern-matching process. *Proc. IXth Int. Cong. on Cybernetics,* Namur, Belgium, pp. 53–65.
4. CAYROL, M., FARRENY, H., and PRADE, H. (1982). Fuzzy pattern matching. *Kybernetes,* **11,** 103–116.
5. CHANG, C. L. (1975). Interpretation and execution of fuzzy programs. In *Fuzzy Sets and Their Applications to Cognitive and Decision Processes* (L. A. Zadeh, K. S. Fu, M. Shimura, and K. Tanaka, eds). Academic, New York, pp. 191–218.
6. CHANG, S. K. (1972). On the execution of fuzzy programs using finite state machines. *IEEE Trans. on Computers,* **21,** 241–253.
7. FARRENY, H., and PRADE, H. (1982). Search methods with imprecise estimates. *Proc. 26th Int. Symposium on General Systems Methodology* (L. Troncale, ed.). Society for General Systems Research, Washington, D.C., pp. 442–446.
8. FARRENY, H., and PRADE, H. (1983). On the problem of identifying an object in a robotics scene from a verbal imprecise description. In *Advanced Software in Robotics* (A. Danthine and M. Géradin, eds.). North-Holland, Amsterdam, pp. 343–351.
9. GOGUEN, J. A. (1975). On fuzzy robot planning. In *Fuzzy Sets and Their Applications to Cognitive and Decision Processes* (L. A. Zadeh, K. S. Fu, K. Tanaka, and M. Shimura, eds.), Academic, New York, pp. 429–447.
10. GONDRAN, M., and MINOUX, M. (1979). *Graphes et Algorithmes.* Eyrolles, Paris.
11. IMAOKA, H., TERANO, T., and SUGENO, M. (1982). Recognition of linguistically instructed path to destination. In *Approximate Reasoning in Decision Analysis* (M. M. Gupta and E. Sanchez, eds.). North-Holland, Amsterdam, pp. 341–350.
12. MCDERMOTT, D., and DAVIS, E. (1984). Planning routes through uncertain territory. *Art. Intell.,* **22,** 107–156.
13. NILSSON, N. (1971). *Problem-Solving Methods in Artificial Intelligence.* McGraw-Hill, New York.
14. NILSSON, N. (1980). *Principles of Artificial Intelligence.* Tioga Publishing, Palo Alto, California.
15. PEARL, J. (1984). *Heuristics—Intelligent Search Strategies for Computer Problem Solving.* Addison–Wesley, Reading, Massachusetts.
16. TANAKA, K., and MIZUMOTO, M. (1973). Fuzzy programs and their execution. In *Fuzzy Sets and Their Applications to Cognitive and Decision Processes* (L. A. Zadeh, K. S. Fu, M. Shimura, and K. Tanaka, eds.). Academic, New York, pp. 41–66.
17. URAGAMI, M., MIZUMOTO, M., and TANAKA, K. (1976). Fuzzy robot controls. *J. Cybernet.,* **6,** 39–64.
18. WINSTON, P. H. (1977). *Artificial Intelligence.* Wiley, New York.
19. YAGER, R. R. (1983). Robot planning with fuzzy sets. *Robotica,* **1,** 41–50.
20. YAGER, R. R. (1986). Paths of least resistance in possibilistic production systems. *Fuzzy Sets Syst.,* **19,** 121–132.
21. ZADEH, L. A. (1968). Fuzzy algorithms. *Inf. Control,* **12,** 94–102.
22. ZADEH, L. A. (1973). Outline of a new approach to the analysis of complex systems and decision processes. *IEEE Trans. Syst., Man Cybernet.,* **3,** 28–44.
23. ZADEH, L. A. (1978). Fuzzy sets as a basis for a theory of possibility. *Fuzzy Sets Syst.,* **1**(1), 3–28.
24. PRADE, H. (1983). Fuzzy programming: Why and how? Some hints and examples. In *Advances in Fuzzy Sets, Possibility Theories and Applications* (P. P. Wang, ed.). Plenum Press, New York, pp. 237–251.
25. DUBOIS, D., and JAULENT, M. C. (1985). Shape understanding via fuzzy models. *Proceedings of the 2nd IFAC/IFIP/IFORS/IEA Conference on the Analysis, Design and Evaluation of Man-Machine Systems,* Varese, Italy, 1985, pp. 302–307.

26. HOGLE, R. A., and BONISSONE,P. P. (1984). A fuzzy algorithm for path selection in autonomous vehicle navigation. *Proc. 23d IEEE Conf. on Decision and Control,* 898–900.
27. FARRENY, H., and PRADE, H. (1987). Uncertainty handling and fuzzy logic control in navigation problems. *International Conference on Intelligent Autonomous Systems* (L. O. Hertzberger and F. C. A. Groen, eds.), North-Holland, Amsterdam, 218–225.

6

Handling Incomplete or Uncertain Data and Vague Queries in Database Applications

One often has to handle data that are far from precise and certain. In fact, the value of an attribute of an object may be completely unknown, incompletely known (i.e., only a subset of possible values of the attribute is known), or uncertain (e.g., a probability or possibility distribution for its value is known). In addition, the attribute may not be applicable to some of the objects being considered and, in certain cases, we may not know whether the value even exists, or whether it is simply not known.

Many approaches to the problem of missing values have been proposed (especially values that are "completely unknown" or "inapplicable")—see Biskup [4], Codd [11], Grant [18]. Grant [19] has considered incompletely known values and Lipski [22, 23] has developed a model for handling problems with incomplete data (incompletely known or completely unknown)—see also Siklóssy and Laurière [36]. Wong [41] proposed a statistical approach to modeling uncertain information in a database. Kunii [47], Tahani [37], and Bose *et al.* [50] have developed methods for dealing with vague requests (represented by fuzzy sets) about precise data and, more recently, Baldwin [2, 3], Buckles and Petry [6, 7, 8] and Umano [38, 39] have introduced various "fuzzifications" of relational algebra with a view to handling nonprobabilistic uncertainty. See also Dockery [51], Montgomery and Ruspini [24], Ruspini [35], and Zemankova and Kandel [48].

In the following we propose a general model (Prade [31], Prade and

Testemale [33]), based on Zadeh's possibility theory [43, 44, 45], which can be used with completely unknown values, incomplete data, and uncertain information, as well as inapplicable attributes. Queries with vague predicates can also be handled. The proposed extension belongs to the relational database model (see Adiba and Delobel [1] and Date [12] for a detailed presentation of basic concepts), as, it may be observed, do most of the approaches mentioned above. The logical approach developed by Bossu and Siegel [5] for disjunctive and negative information tackles a similar problem but in a different framework.

The following paragraph concerns the use of possibility distributions to represent incomplete or uncertain data. Differences and similarities with other approaches based on fuzzy sets are emphasized, and we touch briefly on problems of functional dependence. Then we present the extended relational algebra and the associated query language. The final paragraph is devoted to an example that illustrates the various characteristics of our approach and leads on to some practical problems arising in implementation.

6.1. Representation of Incomplete or Uncertain Data

6.1.1. Representing Data by Means of Possibility Distributions

Let A be an attribute whose domain is D, i.e., D is the set of all possible values for A. Our entire knowledge about the value of A for an object x will be represented by a possibility distribution $\pi_{A(x)}$ on $D \cup \{e\}$, where e is an extraneous element denoting the case where A does not apply to x. In other words, $\pi_{A(x)}$ is a mapping from $D \cup \{E\}$ into $[0, 1]$.

Consider as an example that we want to represent our knowledge about the age of Paul's car. We can distinguish the following situations:

- We do not know whether Paul has a car and, if so, we do not know its age:

$$\pi_{\text{age-of-car(Paul)}}(d) = 1, \qquad \forall\, d \in D \cup \{e\}$$

- It is quite possible that Paul has no car, and at the same time there is a possibility $\lambda > 0$ that he has a car that is more than 5 years old but certainly not a more recent one:

$$\pi_{\text{age-of-car(Paul)}}(d) = \begin{cases} \lambda & \text{if } d \geqslant 5 \\ 0 & \text{if } d < 5 \end{cases}$$

$$\pi_{\text{age-of-car(Paul)}}(e) = 1$$

(the above possibility distribution is the least restrictive one compatible with the available data).

- We are absolutely sure that Paul has no car:

$$\pi_{\text{age-of-car(Paul)}}(e) = 1$$

$$\text{and} = 0 \text{ elsewhere}$$

(This is an instance of the missing value "inapplicable".)

- It is quite possible that Paul has a car, in which case it is new; but there is a nonzero possibility λ that he has no car:

$$\pi_{\text{age-of-car(Paul)}}(e) = \lambda, \qquad \lambda > 0$$

$$\pi_{\text{age-of-car(Paul)}}(d) = \mu_{\text{new}}(d), \qquad \forall\, d \in D$$

where μ_{new} is a membership function representing the vague predicate "new."

- We are quite sure that Paul has a car, but we have no information about its age:

$$\pi_{\text{age-of-car(Paul)}}(e) = 0$$

$$\pi_{\text{age-of-car(Paul)}}(d) = 1, \qquad \forall\, d \in D.$$

(This is an instance of the null value "completely unknown.")

- We are quite sure that Paul has a car and we also have the following incomplete nonfuzzy information: it is between two and four years old:

$$\pi_{\text{age-of-car(Paul)}}(e) = 0$$

$$\pi_{\text{age-of-car(Paul)}}(d) = \begin{cases} 1 & \text{if } d \in [2, 4] \subseteq D \\ 0 & \text{otherwise} \end{cases}$$

- We are quite sure that Paul has a car and we also have the fuzzy data that it is new:

$$\pi_{\text{age-of-car(Paul)}}(e) = 0$$

$$\pi_{\text{age-of-car(Paul)}}(d) = \mu_{\text{new}}(d), \qquad \forall\, d \in D$$

- We are quite sure that Paul has a car and we know its age

exactly—two years:

$$\pi_{\text{age-of-car(Paul)}}(e) = 0$$

$$\pi_{\text{age-of-car(Paul)}}(2) = 1$$

and $= 0$ for all other $d \in D$.

Observe that in each case the possibility distribution has been normalized on $D \cup \{e\}$, which is natural since $D \cup \{e\}$ exhausts all possible alternatives. This approach therefore proposes a single framework for the representation of precise value of an attribute, missing values, incomplete or fuzzy data, and also the case where there is nonzero possibility that an attribute is inapplicable. Unless otherwise stated, we shall suppose below that we always have either $\pi(e) = 0$ or else $\pi(d) = 0,\ \forall\, d \in D$.

In relational databases, entities and associations are represented by relations defined on Cartesian products of domains. In general, an n-ary relation may be imagined as a table with rows and columns; a line (or n-tuple) corresponds to the data about an object and a column corresponds to an attribute. In an ordinary database, only actual elements of the respective domains appear in the cells of a table. In our approach, any possibility distribution normalized on $D \cup \{e\}$ may appear in the cells of the column corresponding to a domain D; thus an extended relation is an ordinary relation defined on a Cartesian product of sets whose elements are possibility distributions (or fuzzy sets, if preferred), rather than a fuzzy relation.

For example, in a database, the entity PERSON may correspond to the following table:

PERSON	NAME	AGE	MARITAL STATUS[a]	SALARY
	. . .	. . .	. . .	. . .
	Paul	30	"Unknown"	About 500
	John	[30, 35]	{1/S, 0.7/D}	About 500
	Martina	Young	M	[100, 300]
	David	15	S	"Inapplicable"
	Frances	About 40	{1/W, 1/D}	[500, 600]

[a] S, single; M, married; D, divorced; W, widowed.

In the above example, some values are represented by possibility distributions on continuous domains and others on discrete domains (the notation $\{1/S, 0.7/D\}$ means that there is possibility 1 that the person is single, possibility 0.7 that the person is divorced, and zero possibility for the other alternatives). "Young," "about 40," and "about 500" are names of fuzzy sets.

If the value a of a single-valued attribute for an object x is known exactly, we can deduce logically that A can take no value other than a for x. The situation is different when we only know possibility distributions (not singleton) restricting the possible values of an attribute for an object x; in this case, we cannot always declare that an assertion concerning the value of an object is definitely true or definitely false. All the available data are explicitly represented by the possibility distributions.

N.B. In this chapter, we only consider single-value attributes. Thus when the value appearing in a cell of the relation is a subset (fuzzy or not) of the domain, representing a possibility distribution, the elements of this subset are *mutually exclusive* possible values of the associated attribute; the degree of membership of an element measures the possibility that the element is the exact value of the attribute. In the representation used in this chapter, we do not model linked values. For example, if we know that two people have the same age, and that they are young, but we do not know their age exactly, then it is not enough to represent merely the fact that each is young, for, in a subsequent search for people of the same age, these two will be recognized as being possibly but not necessarily of the same age (see Section 6.2). To represent such a link, we can use the indexing method proposed by Siklóssy and Laurière [36]. Prade and Testemale [34] study in detail how to use possibility theory to handle imprecise information about multivalued attributes (i.e., which can simultaneously take several values as, for example, the languages spoken by someone); they also study representation and use of possible imprecise constraints.   □

6.1.2. Differences and Similarities with Other Fuzzy Approaches

The idea of using possibility distributions or related concepts for modeling incomplete or fuzzy knowledge in databases has already been used by Buckles and Petry [6, 7, 8], Umano [38, 39], and Prade [27]. Tahani [37] uses fuzzy terms uniquely for formulating fuzzy queries about precise data; the reply to a query consists of a fuzzy set of data. Buckles and Petry introduce a fuzzy similarity relation attached to each domain, so as to model the extent to which two elements of the domain may be interchanged. Thus the possible values of an attribute are represented by (nonfuzzy) sets of elements which are interchangeable according to the similarity relation with a fixed threshold depending on the domain. Umano [38], on the other hand, proposes a model based explicitly on the concept of possibility distribution. In our approach we generalize this kind of representation by introducing an extraneous element to cover cases where there is nonzero possibility of an inapplicable attribute. However, our method of query evaluation (see Section 6.2) is different; it is based on the dual concepts of possibility and necessity measures, while Umano's [38] is based rather on an ad hoc logic.

Another way to represent fuzzy knowledge consists of associating a "fuzzy truth-value" (i.e., a number in [0, 1]) to each item of knowledge about an object; see Le Faivre [21], Haar [20], Philips, Beaumont, and Richardson [26], Freksa [16] (who uses instead linguistic truth values represented by a possibility distribution on [0, 1]), Baldwin [2, 3], and Umano [39] (who also uses truth values defined by possibility distributions on [0, 1]). In a relational database, this mode of representation gives rise to n-tuples consisting solely of elements from the attribute domains, each n-tuple being evaluated by a possibility distribution on [0, 1] (perhaps reduced to a single value between 0 and 1). Such a representation corresponds to a fuzzy relation, and contrasts with the approach based on possibility distributions where an n-tuple is an ordered collection of possibly fuzzy attribute values, without any truth value.

In a more recent article, Umano [39] combines the two representations and obtains n-tuples that are ordered collections of fuzzy sets which, in addition, are evaluated by a "degree of truth" (perhaps defined by a possibility distribution on [0, 1]). However, no indication is given of how to interpret these degrees of truth in terms of possibility theory, and their meaning must therefore be found empirically, as also must the way they are to be used in evaluating queries. Zadeh [44], moreover (and see also Chapter 4, Section 4.1), has proposed an approach to assigning a linguistic truth value to a proposition partially specifying the value of an attribute, which reduces this to specifying a simple possibility distribution on possible attribute values; this is apparently not the approach chosen by Umano [39]. It is interesting all the same to note that when all the n-tuples have value 1, Umano's method [39] yields, as answer to a query, a fuzzy set whose degrees of membership are themselves possibility distributions on [0, 1], and from it can be deduced the result that would be yielded by our approach as answer to the same query. Nonetheless it seems to us to be better to work directly, and explicitly, in terms of possibility and necessity.

Baldwin [2] also uses a mixed approach: a representation based on possibility distributions for entities (with fuzzy sets to restrict the possible attribute values) and a representation based on truth values to model fuzzy associations between entities. However, Baldwin [2] maintains the homogeneity of his representation by attaching a truth value of 1 to each n-tuple belonging to a relation defining an entity. On the other hand, it seems possible to transform the fuzzy relation R representing an association into an ordinary relation whose n-tuples are collections of fuzzy subsets, as suggested by the following example:

LIKES	NAME1	NAME2	R
	John	Francis	0.8
	John	Mary	1
	James	Frederick	0.5
	Frederick	David	0.5
	David	Mary	0.2
	Mary	John	0.8

is transformed into

SENTIMENT	NAME1	NAME2	TYPE
	John	Francis	h
	John	Mary	i
	James	Frederick	g
	Frederick	David	g
	David	Mary	f
	Mary	John	h

where the domain of TYPE is $\{a, b, c, d, e, f, g, h, i\}$, each letter denoting a standard level of sentiment between two people; for example a corresponds to "deadly hatred", e to "indifference" and i to "deep friendship." LIKES is here considered as a fuzzy subset of the domain of TYPE whose membership function has been obtained by exemplification, i.e., for the n-tuple x, $\mu_R(x) = \mu_{\text{LIKES}}(\text{TYPE}(x))$. More generally, it is clear, any fuzzy subset of the domain of TYPE can be a value in the TYPE column; in this way we can represent incomplete knowledge about the feelings between two people, which is more difficult in a model using truth values.

Remark. In some situations, it may seem natural to use probability distributions to model uncertain information—see Wong [41]. As pointed out in Chapter 1, Section 1.6.2.2, one can define a bijection between a probability distribution $p = (p_1, \ldots, p_n)$ and a possibility distribution

$\pi = (\pi_1, \ldots, \pi_n)$ [formulas (1.57) and (1.58)]. Thus a "possibilistic" interpretation of frequency histograms can be given which supplements the usual probabilistic interpretation. In other words, statistical data may also be represented by possibility distributions.   □

6.1.3. Dependencies and Possibilistic Information

In databases containing precise and certain data, a functional dependence $A \to B$ between attributes A and B is expressed, for each pair (x, y) of n-tuples, by the following implication:

$$\text{If} \quad A(x) = A(y) \quad \text{then} \quad B(x) = B(y) \tag{6.1}$$

where $A(x)$ denotes the value of attribute A for the object corresponding to the n-tuple x. A functional dependence corresponds to a function from the domain of A to that of B. Representation of functional dependence models constraints existing in reality which must be taken into account in order that modifications to the data should not lead to contradictions.

When the database contains partial (and imprecise) data, we cannot use (6.1) directly. In fact, equality of the possibility distributions restricting the possible values of attribute A for x and y does not imply equality of the possibility distributions restricting the possible values of B for x and y, except when the distribution associated with $A(x)$ and $A(y)$ is reduced to a singleton, or we have supplementary information asserting equality of the values of A for x and for y; only in these two cases may we conclude that the values of B for x and y are equal, and therefore that the possibility distributions for $B(x)$ and $B(y)$ are equal.

On the other hand, it is interesting to "fuzzify" the concept of dependence in order to deal with situations like "age approximately determines salary." More generally, if the values of attribute A for x and y are equal, we wish to express the notion that the values of B for x and y are not too different; we can model this idea by means of a fuzzy proximity relation P on the domain D_B of attribute B, i.e., a fuzzy reflexive and symmetric relation P:

$$\forall d, \quad \mu_P(d, d) = 1, \quad \forall d, \quad \forall d', \quad \mu_P(d, d') = \mu_P(d', d)$$

Thus we obtain the fuzzy version of (6.1):

$$\text{If } A(x) = A(y), \quad \text{then } \mu_P(B(x), B(y)) \geq \alpha \tag{6.2}$$

where α is a given threshold.

Next, bearing in mind that $B(x)$ and $B(y)$ are only known in terms of possibility distributions, we obtain

$$\text{If } A(x) = A(y), \quad \text{then } \Pi(B(x) \approx_P B(y)) \geq \alpha \quad (6.3)$$

where $\Pi(B(x) \approx_P B(y))$ is the possibility that the values of B for x and y are approximately equal in the P sense; this possibility is equal to

$$\Pi(B(x) \approx_P B(y)) = \sup_{(v,w) \in D_B \times D_B} \min(\mu_P(v, w), \pi_{B(x)}(v), \pi_{B(y)}(w)) \quad (6.4)$$

where $\pi_{B(x)}$ is the possibility distribution restricting the possible values of attribute B for x, and μ_P is the membership function of the fuzzy proximity relation P. See [34] for a more systematic treatment of such constraints.

6.2. The Extended Relational Algebra and The Corresponding Query Language

6.2.1. Generalization of Θ-Selection

6.2.1.1. Principal Characteristics

Θ-selection consists of retrieving the n-tuples of a relation whose components satisfy a given condition (atomic or compound); this operation plays an important role in a query language. Θ-selection applied to a relation R with a condition $\mathscr{C}$ is denoted by $\sigma(R; \mathscr{C})$. A compound condition $\mathscr{C}$ is constructed from atomic conditions by logical connectives (negation, conjunction, disjunction). We distinguish two kinds of atomic condition: conditions of the form "$A \Theta B$" with two attributes A and B and a comparator Θ, and on the other hand conditions of the form "$A \Theta a$" with an attribute A and a constant a which may or may not be fuzzy. In ordinary relational algebra, Θ is one of the comparators $=, \neq, >, \geq, <, \leq$; in our approach, we can represent a comparison (fuzzy or not) by a membership function μ_Θ defined on a Cartesian product of two domains and taking values in $[0, 1]$, so that fuzzy comparators like "approximately equal" and "much greater than" can be modeled.

The use of possibility measures to restrict the possible values of each attribute induces a possibility measure and a necessity measure to evaluate the fact that an n-tuple satisfies a given condition. Recall that possibility and necessity measures constructed from a possibility distribution π are defined, for a fuzzy event F, by formulas (1.63) and (1.66)

$$\Pi(F) = \sup_{\omega \in \Omega} \min(\mu_F(\omega), \pi(\omega)) \tag{6.5}$$

$$N(F) = \inf_{\omega \in \Omega} \max(\mu_F(\omega), 1 - \pi(\omega)) \tag{6.6}$$

Here $\mathscr{C}$ plays the role of F, and π is built from tuple x of R.

The result of a Θ-selection $\sigma(R; \mathscr{C})$ is therefore made up of two fuzzy sets: the set of n-tuples of R *possibly* satisfying the condition $\mathscr{C}$, and the set *necessarily* satisfying $\mathscr{C}$, the degree of membership corresponding to a measure of, respectively, possibility or necessity. For partial nonfuzzy data, Lipski [22, 23] and Narin'Yani [25] have also distinguished between the n-tuples that surely satisfy a query and those that possibly satisfy the same query.

We write

$$\sigma(R; \mathscr{C}) = (\sigma\Pi(R; \mathscr{C}), \sigma N(R; \mathscr{C})) \tag{6.7}$$

to show that the result of a Θ-selection is a pair of two fuzzy sets. Since the possibility of an event (fuzzy or not) is always greater than or equal to its necessity, we obtain for each n-tuple x of the relation R

$$\mu_{\sigma\Pi(R;\mathscr{C})}(x) \geqslant \mu_{\sigma N(R;\mathscr{C})}(x) \tag{6.8}$$

where μ denotes a membership function; (6.8) expresses the fuzzy-set inclusion $\sigma N(R; \mathscr{C}) \subseteq \sigma\Pi(R; \mathscr{C})$.

6.2.1.2. Compound Conditions

Evaluation of a compound condition can be reduced to evaluation of its atomic conditions by virtue of the following results:

$$\sigma(R; \neg\mathscr{C}) = (\overline{\sigma N(R; \mathscr{C})}, \overline{\sigma\Pi(R; \mathscr{C})}) \tag{6.9}$$

$$\sigma(R; \mathscr{C}_1 \vee \mathscr{C}_2) = (\sigma\Pi(R; \mathscr{C}_1) \cup \sigma\Pi(R; \mathscr{C}_2), \sigma N(R; \mathscr{C}_1) \cup \sigma N(R; \mathscr{C}_2)) \tag{6.10}$$

$$\sigma(R; \mathscr{C}_1 \wedge \mathscr{C}_2) = (\sigma\Pi(R; \mathscr{C}_1) \cap \sigma\Pi(R; \mathscr{C}_2), \sigma N(R; \mathscr{C}_1) \cap \sigma N(R; \mathscr{C}_2)) \tag{6.11}$$

where $\overline{\sigma N(R; \mathscr{C})}$ and $\overline{\sigma\Pi(R; \mathscr{C})}$ denote the complements of $\sigma N(R; \mathscr{C})$

and $\sigma\Pi(R; \mathscr{C})$, defined as 1 minus the degree of membership, and union and intersection are, respectively, defined by the max and min operators. Equation (6.9) is a direct consequence of $N(F) = 1 - \Pi(\bar{F})$. The formulas (6.10) and (6.11) only hold if the attributes occurring in conditions $\mathscr{C}_1$ and $\mathscr{C}_2$ are noninteractive (two attributes A and B are called noninteractive if, for each n-tuple, the value of one does not depend on the value of the other); see Chapter 1, Section 1.8.

6.2.1.3. Atomic Conditions with One Attribute and a Constant

Consider first atomic conditions of the form "$A \ominus a$," where A is an attribute, $\ominus$ a comparator represented by its membership function $\mu_\ominus$, and a a constant likewise represented by its membership function μ_a. "Age much greater than 30 years" and "Age equals young" (corresponding to the query "Find all the people who are young") are examples of such conditions.

The possibility that the value of attribute A for object x belongs to the set of elements that are in relation $\ominus$ with at least one element of a is given by

$$\Pi(a \circ \ominus \mid A(x)) = \sup_{d \in D} \min(\mu_{a \circ \ominus}(d), \pi_{A(x)}(d)) \tag{6.12}$$

with

$$\mu_{a \circ \ominus}(d) = \sup_{d' \in D} \min(\mu_\ominus(d, d'), \mu_a(d')) \tag{6.13}$$

where D is the domain of attribute A, $\mu_\ominus$ is the membership function of a relation (fuzzy or not) defined on $D \times D$, and $\pi_{A(x)}$ is the possibility distribution restricting the possible values of attribute A for x with the supplementary condition $\pi_{A(x)}(e) = 0$ (e is the extraneous element introduced in Section 6.1.1).

The necessity of the same event is given by

$$N(a \circ \ominus \mid A(x)) = \inf_{d \in D} \max(\mu_{a \circ \ominus}(d), 1 - \pi_{A(x)}(d)) \tag{6.14}$$

N.B. In this chapter, the symbol $\mid$ does not denote conditioning, but serves to separate the "query" from the item in the database.   □

$\Pi(a \circ \ominus \mid A(x))$ and $N(a \circ \ominus \mid A(x))$, respectively, define the degree of membership of x in the set of n-tuples that possibly or necessarily satisfy the condition $A \ominus a$, *i.e.* ,

$$\mu_{\sigma\Pi(R; A \ominus a)}(x) = \Pi(a \circ \Theta \,|\, A(x)) \tag{6.15}$$

$$\mu_{\sigma N(R; A \ominus a)}(x) = N(a \circ \Theta \,|\, A(x)) \tag{6.16}$$

Formula (6.13) says that $a \circ \Theta$ is the fuzzy set of elements of D that are in relation Θ with at least one element of a; (6.13) extends to fuzzy sets the following expression, which holds for nonfuzzy sets:

$$a \circ \Theta = \{d \in D \,|\, a \cap \Theta_d \neq \varnothing\} \tag{6.17}$$

where Θ_d is the set of elements of d which are in relation Θ with d. Note that if Θ is reflexive [i.e., $\forall\, d \in D$, $\mu_\Theta(d, d) = 1$], then $\mu_{a \circ \Theta} \geqslant \mu_a$, which is to say that $a \circ \Theta \supseteq a$. Formulas (6.12) and (6.14) follow directly from (6.5) and (6.6). If a can be decomposed into the union of two fuzzy sets b and c, union being defined by "max," we have

$$a \circ \Theta = (b \cup c) \circ \Theta = (b \circ \Theta) \cup (c \circ \Theta) \tag{6.18}$$

There is no analogous formula for the intersection. It can be verified that

$$\Pi((b \cup c) \circ \Theta \,|\, A(x)) = \max(\Pi(b \circ \Theta \,|\, A(x)), \Pi(c \circ \Theta \,|\, A(x))) \tag{6.19}$$

But we only have

$$N((b \cup c) \circ \Theta \,|\, A(x)) \geqslant \max(N(b \circ \Theta \,|\, A(x)), N(c \circ \Theta \,|\, A(x))) \tag{6.20}$$

Thus, to calculate (6.14), a cannot be decomposed into the union of two subsets. Moreover, there are situations where the disjunction of two predicates cannot be satisfactorily represented as the union of two fuzzy sets, in which case the convex hull of the union gives a more suitable representation. The convex hull $\hat{a}$ of a fuzzy set a defined on a totally ordered universe D is defined by

$$\mu_{\hat{a}}(d) = \sup_{d' \leqslant d \leqslant d''} \min(\mu_a(d'), \mu_a(d'')) \tag{6.21}$$

Note that $\mu_{\hat{a}} \geqslant \mu_a$, i.e. $\hat{a} \supseteq a$; the convex hull of the union of two fuzzy sets b and c is shown in Figure 6.1. Thus the convex hull of the fuzzy set representing "medium or large" includes all elements "between" "medium" and "large," with a degree of membership equal to 1.

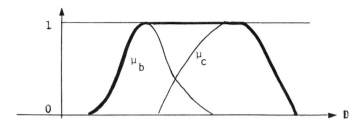

Figure 6.1. Convex hull of two fuzzy sets. ——, $\mu_b \hat{\cup} c$.

N.B. When $\pi_{A(x)}(e) \neq 0$, i.e., when there is nonzero possibility that attribute A is inapplicable to x, then (6.15) and (6.16) become

$$\mu_{\sigma\Pi(R; A\ominus a)}(x) = \Pi(a \circ \Theta \,|\, A^*(x)) \tag{6.22}$$

$$\mu_{\sigma N(R; A\ominus a)}(x) = 1 - \Pi(\overline{a \circ \Theta} \cup \{e\} \,|\, A(x)) \tag{6.23}$$

where the asterisk indicates that the restriction of $\pi_{A(x)}$ to D is used in (6.22) [and not $\pi_{A(x)}$ defined on $D \cup \{e\}$ as in (6.23)]. To justify (6.23), observe that the necessity that x satisfies condition $A \ominus a$ corresponds to the impossibility of the event "A is inapplicable to x or else the value $A(x)$ does not satisfy the condition $A \circ a$"; the possibility measure in (6.23) is calculated by replacing $a \circ \Theta$ by $\overline{a \circ \Theta} \cup \{e\}$, and D by $D \cup \{e\}$ in (6.12); the overbar here refers to complementation in D. ☐

Finally, consider the case where Θ is "equality"; $a \circ \Theta = a$, and formulas (6.12) and (6.14), respectively, express the possibility and the necessity that the value $A(x)$ belongs to a. More generally, it must be noted that if a is a singleton $\{d_0\}$, then "$A(x)$ is in relation Θ with a" has the usual meaning that "$A(x)$ is in relation Θ with d_0" while in the contrary case "$A(x)$ is in relation Θ with a" means, in our approach, that "$\exists\, d \in a$ such that $A(x)$ is in relation Θ with d."

6.2.1.4. Atomic Conditions with Two Attributes

We now consider atomic conditions of the form "$A \ominus B$," where A and B are distinct attributes with the same domain, and Θ is a comparator (fuzzy or not) represented by its membership function μ_Θ. In the following, we suppose that attributes are always applicable. Such conditions hold in the following examples: "Find all students whose ability in mathematics is approximately equal to their ability in physics" or "Find all students whose ability in mathematics is greater than their ability in physics."

The possibility that the value of attribute A for object x is in relation Θ with the value of attribute B for the same object x is given by

$$\Pi(\Theta \mid (A(x), B(x))) = \sup_{(d,d')\in D\times D} \min(\mu_\Theta(d, d'), \pi_{(A(x),B(x))}(d, d'))$$

$$(6.24)$$

where $\pi_{(A(x),B(x))}$ is the possibility distribution restricting the possible values of the pair (A, B) for x. When the attributes A and B are noninteractive, $\pi_{(A(x),B(x))}$ can be decomposed into

$$\forall (d, d') \in D \times D, \ \pi_{(A(x),B(x))}(d, d') = \min(\pi_{A(x)}(d), \pi_{B(x)}(d')) \quad (6.25)$$

(see Chapter 1, Section 1.8). Equation (6.24) then becomes

$$\Pi(\Theta \mid (A(x), B(x))) = \sup_{(d,d')\in D\times D} \min(\mu_\Theta(d, d'), \pi_{A(x)}(d), \pi_{B(x)}(d'))$$

$$(6.26)$$

The necessity that the value of A for x is in relation Θ with the value of B for x is given by

$$N(\Theta \mid (A(x), B(x))) = \inf_{(d,d')\in D\times D} \max(\mu_\Theta(d, d'), 1 - \pi_{(A(x),B(x))}(d, d'))$$

$$(6.27)$$

which for noninteractive attributes becomes

$$N(\Theta \mid (A(x), B(x)))$$

$$= \inf_{(d,d')\in D\times D} \max(\mu_\Theta(d, d'), 1 - \pi_{A(x)}(d), 1 - \pi_{B(x)}(d')) \quad (6.28)$$

Note that formulas (6.24) and (6.27) follow directly from (6.5) and (6.6). Moreover, (6.26) and (6.28) are straightforward extensions of formulas (3.25) and (3.28), derived for fuzzy number comparisons.

$\Pi(\Theta \mid (A(x), B(x)))$ and $N(\Theta \mid (A(x), B(x)))$ define the degrees of membership of x in the set of n-tuples which possibly or necessarily satisfy the condition A⊖B, i.e.,

$$\mu_{\sigma\Pi(R; A\ominus B)}(x) = \Pi(\Theta \mid (A(x), B(x))) \quad (6.29)$$

$$\mu_{\sigma N(R; A\ominus B)}(x) = N(\Theta \mid (A(x), B(x))) \quad (6.30)$$

For symmetric comparators (equality, approximate equality) it can be verified that the expressions (6.26) and (6.28) are themselves symmetric in A and B. More precisely, for the comparison $A = B$, (6.28) evaluates the necessity that the values of the attributes A and B for x are equal, but does not evaluate to what extent the possibility distributions $\pi_{A(x)}$ and $\pi_{B(x)}$ are equal . In fact, (6.28) returns a nonzero value only if $\exists\, d_0 \in D$ such that $\pi_{A(x)}(d_0) = \pi_{B(x)}(d_0) = 1$ and $\exists\, \alpha$ such that $\forall\, d \neq d_0$, $\pi_{A(x)}(d) < \alpha$ and $\pi_{B(x)}(d) < \alpha$ with $\alpha < 1$; the value given by (6.28) equals 1 only if the values of the attributes are precisely known and are equal.

N.B. Note that a query like "find all the people whose age is approximately equal to the age of Paul" is handled in terms of the approach used for conditions of the form $A \ominus B$, though the associated condition may seem to be of the type $A \ominus a$, where a represents the information concerning the age of Paul. In fact, we must compare the age of x with the age of Paul, and not evaluate to what extent the age of x belongs to the fuzzy set of values approximately equal to one of the possible ages of Paul. More generally, we may verify that

$$N(a \circ \ominus \mid A(x)) \geq N(\ominus \mid (A(x), a)) \qquad (6.31)$$

and

$$\Pi(a \circ \ominus \mid A(x)) = \Pi(\ominus \mid (A(x), a)) \qquad (6.32)$$

where $a = A(x_0)$; except in certain pathological cases, (6.31) is a strict inequality; in particular, equality holds if a is a singleton. $\qquad\square$

6.2.1.5. Tolerance

The purpose of this paragraph is twofold: we first examine the consequences of possible imprecision for the membership functions which define possibility distributions; then we study how to take account of variations in the content of the restrictive condition $\mathscr{C}$, and their consequences for the result of the selection $\sigma(R; \mathscr{C})$.

As observed in [32], if ε is the maximum size of the possible absolute error in the values of the possibility distributions at each point of their respective domains, this error cannot change the values of derived possibility and necessity distributions by more than ε. In probability theory, however, errors in distributions may have greater repercussions.

A condition $\mathscr{C}$ in a selection $\sigma(R; \mathscr{C})$ involves one or several comparators, according as it is atomic or compound. We can make the condition more or less restrictive by modifying any of these comparators. A comparator $\ominus$ can be relaxed (e.g., changing strict equality into an approximate equality) by taking the composition of the comparator with a tolerance relation T (fuzzy or not):

$$\mu_{\Theta \circ T}(d, d') = \sup_{d'' \in D} \min(\mu_\Theta(d, d''), \mu_T(d'', d')) \qquad (6.33)$$

where μ_T is the membership function of the tolerance relation, which must be reflexive and symmetric, i.e.,

$$\forall d \in D, \qquad \mu_T(d, d) = 1$$

$$\forall (d'', d') \in D \times D, \qquad \mu_T(d'', d') = \mu_T(d', d'')$$

Note that in the case of a condition $A \ominus a$, introducing a tolerance relation T amounts to augmenting the set of elements that are in relation Θ with at least one element a, since we have

$$a \circ (\Theta \circ T) = (a \circ \Theta) \circ T \qquad (6.34)$$

where $\circ$ is the sup–min composition. In practice, computing the sup–min composition (6.33) comes down to addition of fuzzy numbers (see the remark at the end of Section 2.3.2 of Chapter 2).

6.2.2. Cartesian Product, Θ-Join, and Projection

The Cartesian product of two extended relations R and S (whose elements are ordered collections of possibility distributions) is defined in the usual way since R and S are ordinary relations on the Cartesian product of the sets of fuzzy subsets of the domains of the respective attributes.

Consider the two extended relations

R	A_1	A_2	A_3
	u_1	u_2	u_3
	v_1	v_2	v_3
	w_1	w_2	w_3

and

S	A_3	A_4
	x_3	x_4
	y_3	y_4
	z_3	z_4
	t_3	t_4

Their Cartesian product is given by

$R \times S$	A_1	A_2	A_3	A_3	A_4
	u_1	u_2	u_3	x_3	x_4
	...	...	...	...	...
	u_1	u_2	u_3	t_3	t_4
	v_1	v_2	v_3	x_3	x_4
	...	...	...	...	...
	w_1	w_2	w_3	t_3	t_4

In databases containing only exact values and using only nonfuzzy comparators, the Θ-join of the relations R and S corresponds to concatenation and to selection of pairs of n-tuples such that a comparison Θ between some of their components is satisfied. It is in fact a Θ-selection on the Cartesian product $R \times S$.

An extended Θ-join can be defined as an extended Θ-selection on a Cartesian product. In the above example, the extended equijoin (Θ is equality) corresponds to the concatenation of pairs of n-tuples having possibly and necessarily the same value on the attribute A_3 of R and S, and is given by

$$\sigma(R \times S; A_3^{(R)} = A_3^{(S)})$$

$$= (\sigma\Pi(R \times S; A_3^{(R)} = A_3^{(S)}), \sigma N(R \times S; A_3^{(R)} = A_3^{(S)}))$$

and, more explicitly, we obtain the fuzzy relations

$\sigma\Pi(R \times S; A_3^{(R)} = A_3^{(S)})$	A_1	A_2	A_3	A_3	A_4	Π
	u_1	u_2	u_3	x_3	x_4	$\Pi(= \|(u_3, x_3))$
	...	...	...	...	...	...
	u_1	u_2	u_3	t_3	t_4	$\Pi(= \|(u_3, t_3))$
	v_1	v_2	v_3	x_3	x_4	$\Pi(= \|(v_3, x_3))$
	...	...	...	...	...	...
	w_1	w_2	w_3	t_3	t_4	$\Pi(= \|(w_3, t_3))$

and, similarly, $\sigma N(R \times S; A_3^{(R)} = A_3^{(S)})$. Projecting the relation $\sigma\Pi(R \times S; A_3^{(R)} = A_3^{(S)})$ onto the attribute A_1, for example, and the special column Π, we obtain

A_1	Π
u_1	$\Pi(= \|(u_3, x_3))$
...	...
u_1	$\Pi(= \|(u_3, t_3))$
v_1	$\Pi(= \|(v_3, x_3))$
...	...
w_1	$\Pi(= \|(w_3, t_3))$

Note that there may be several nonzero degrees of possibility for the same A_1 component: to complete the projection one retains solely the maximum of the degrees of possibility for each A_1 component. We have in fact calculated the projection of the fuzzy relation $\sigma\Pi(R \times S; A_3^{(R)} = A_3^{(S)})$ (see Chapter 1, Section 1.8).

Suppose that in the above example, A_1, A_2, A_3, and A_4 represent respectively NAME, HEIGHT, AGE, and SALARY; we then obtain the names of persons having a "good" salary as the result of the following operation:

$$\text{PROJ}_{\text{NAME}}(\sigma(R \times S; \text{AGE}^{(R)} = \text{AGE}^{(S)} \wedge \text{SALARY} = \text{``good''})) \quad \text{(A)}$$

where the projection is carried out separately on the relations $\sigma\Pi$ and σN in the way described above. The aim of the following commentary is to explain the precise sense of the expression (A):

From line i, $(n(i), h(i), a_R(i))$, of relation R and line j, $(a_S(j), s(j))$ of relation S, we construct a line of $R \times S$; this line belongs to the part $\sigma\Pi$ of the selection calculated in (A) with degree of membership

$$\min(\Pi(= \|(a_R(i), a_S(j))), \Pi(\text{``good''} \| s(j))) \quad \text{(B)}$$

and belongs to the part σN of the selection with degree

$$\min(N(= \|(a_R(i), a_S(j))), N(\text{``good''} \| s(j))) \quad \text{(C)}$$

Next, the value $n(i)$ of the attribute NAME belongs to the projection of the part $\sigma\Pi$ of (A) with degree

$$\sup_j \min(\Pi(= \|(a_R(i), a_S(j))), \Pi(\text{``good''} \| s(j))) \quad \text{(D)}$$

and to the projection of the part σN of (A) with degree

$$\sup_j \min(N(= \|(a_R(i), a_S(j))), N(\text{``good''} \| s(j))) \quad \text{(E)}$$

In expression (C), we calculate the necessity that the two variables age(i) and age(j) should be equal, the possible values of these variables being restricted by $\mu_{a_R(i)}$ and $\mu_{a_S(j)}$; this evaluation agrees entirely with the following interpretation of S. A line of S corresponds to an imprecise description, in terms of possibility distributions, of a pair of *single-valued* attributes. That is, in our example, we define a function which assigns to *just one* of the ages more-or-less belonging to $a_S(j)$ a salary-value restricted by $s(j)$.

We can give a different interpretation of a line of S: to *every* value of age belonging to $a_S(j)$ will be associated a salary-value restricted by $s(j)$. Note that this interpretation does not agree with the general approach presented in this chapter, since $a_S(j)$ would then be considered as a (possibly fuzzy) set of values and not as a fuzzy set of possible mutually exclusive values (i.e., a possibly distribution). It would, however, be feasible to incorporate this interpretation within the framework we propose, by replacing $N(= | (a_R(i), a_S(j)))$ by $N(a_S(j) | a_R(i))$ in expressions (C) and (E). [Recall that $\Pi(= | (a_R(i), a_S(j))) = \Pi(a_S(i) | a_R(j)).$] Moreover, by expanding expression (D) we obtain

$$\sup_{v} \min\left(\sup_{u} \min\left(\mu_{a_R(i)}(u), \sup_{j} \mu_{a_S(j) \times s(j)}(u, v) \right), \mu_{\text{good}}(v) \right)$$

$$= \Pi\left(\text{``good''} \mid a_R(i) \circ \bigcup_{j} [a_S(j) \times s(j)] \right)$$

that is, the possibility that the salary of the person whose name is in line i of R should be "good." The calculation of the possibility distribution restricting possible salary values agrees with the usual schema of approximate reasoning (see Chapter 4, Section 4.3).

6.2.3. Union and Intersection—Redundancy

As with ordinary relations, two extended relations are called compatible if there is a one-to-one correspondence (bijection) between their attribute sets such that corresponding attributes are defined on the same domain (Codd [11]). The union of two compatible relations R and S corresponds to the usual union defined on subsets of Cartesian products of sets. After making the union, n-tuples considered redundant must be eliminated. In ordinary relational algebra, two n-tuples are redundant if they are identical; the requirement of strict identity may seem too stringent for comparison of possibility distributions since they are generally obtained by approximate identification. We meet a similar problem in the definition of intersection: an n-tuple belongs to $R \cap S$ if it is redundant compared with an n-tuple of R and also with one of S.

We can define approximate equality of two possibility distributions π and π' by

$$\sup_{d \in D} |\pi(d) - \pi'(d)| \leq \varepsilon_D \qquad (6.35)$$

where ε_D is a domain-dependent threshold. As with any approximate equality relation, (6.35) does not define a transitive relation. An extended relation is redundancy-free when no two n-tuples can be found that are approximately equal in each component, in the sense of (6.35); but since (6.35) does not define an equivalence, there is no unique way to eliminate redundancies in a given relation. However, the different redundancy-free relations that may be obtained are similar to each other and one may be used instead of another without affecting responses to queries significantly (compared with ε_D).

More generally, the database reduction problem is addressed in a recent paper (Prade and Testemale [46]). See also Buckles and Petry [6, 7] for a different representation framework.

6.2.4. Queries Employing Other Operations

6.2.4.1. "Yes–No" Queries

Essentially two kinds of query can be put to a database: queries that seek a specified set of objects and queries that seek a yes-or-no answer. In the above paragraphs we have discussed the first kind of query. Now we briefly consider the problem of yes–no queries with incomplete information. Here it may not be possible to reply definitely yes or no, but we can evaluate the possibility and the necessity that the answer is yes; the possibility or necessity that the answer is no is 1 minus the necessity or possibility, respectively, that the answer is yes. Thus the results of the preceding paragraphs can be applied directly to such questions as "Is it true that the value of attribute A for x_0 is in relation Θ with a?" or "Is it true that the value of attribute A for x_0 is in relation Θ with the value of attribute B for x_0?" as well as analogous questions involving compound conditions. Moreover, our approach admits questions with quantifiers (fuzzy or not) such as "Is it true that *most* of the objects for which the value of attribute A is in relation Θ with a have a value of B that is in relation Θ' with b?" (see Prade [31] and Kacprzyk [49]).

6.2.4.2. Cardinality

When the values of an attribute are known only as possibility distributions, one can calculate the possibility distributions π_{card} restricting the possible values for the number of n-tuples for which the value of attribute A is in relation Θ with a. It can be shown (Prade [31]) that this possibility distribution equals $\forall\, n \in N$,

$$\pi_{card(A\,\Theta\,a)}(n) = \min\left(\max_{\substack{S\subseteq T\\|S|=n}} \min_{x\in S} \Pi(a\circ\Theta\,|\,A(x)),\ \min_{\substack{S\subseteq T\\|S|=n+1}} \max_{x\in S} \Pi(\overline{a\circ\Theta}\,|\,A(x))\right)$$

$$(6.36)$$

where T is the set of n-tuples of the relation considered. We can therefore reply to a query bearing on cardinality in terms of a possibility distribution. Further, results concerning relations of fuzzy order between fuzzy numbers allow queries concerning comparison of cardinals to be handled (see Dubois and Prade [14], and Chapter 3, Section 3.2). Despite its impressive look, (6.36) remains easy to compute in practice.

6.3. Example

6.3.1. Representation of Data

Consider an example of data on students concerning their attainments in different subjects, their age, and their degrees of liking for each other. The database contains the following two relations:

PERSON NAME	AGE	M1	M2	P1	P2
Tom	Young	15	Fairly good	Unknown	[14, 16]
David	20	Fairly bad	Good	Fairly good	N/A
Bob	22	Bad to very bad	[10, 12]	[13, 20]	Good
Jane	About 21	Fairly good	Very good	14	[10, 12]
Jill	Young	About 10	Fairly good	Good	About 12
Joe	About 24	[14, 16]	Very good	Good	15
Jack	[22, 25]	Unknown	Bad	About 13	Fairly good

where M1 and P1 are the marks in mathematics and physics in the first term, and M2 and P2 are the marks for the second term.

FEELING	NAME1	NAME2	TYPE
	David	Tom	Camaraderie
	Jill	Jane	Friendship
	Jack	Joe	Great friendship
	Jane	Bob	Friendship

where, for a given triplet, column 1 gives the name of a person who has the sentiment in column 3 toward the person named in column 2.

In the first relation, the domain of the attributes M1, P1, M2, P2 is the interval $[0, 20]$ and the domain of AGE is $[15, 25]$ (these are continuous domains). Many kinds of attribute value appear in this relation: precise values (e.g.,15), intervals (e.g., $[14, 16]$, $[13, 20]$), fuzzy values (e.g., "good," "about 10," "bad-to-very-bad") and missing values ("unknown" and "N/A" = "inapplicable"). Once the attainment is defined (i.e., the student has finished the course), imprecise attribute values may correspond to partial (incomplete or fuzzy) knowledge of attainment in a given course, to an overall estimate of attainment, or to a possibilistic representation of the histogram of (precise) marks obtained by the student in the course of the term. In the last case, the associated possibility distribution corresponds to the aggregate of marks that are individually more or less compatible with the attainments of the student.

In the second relation, representing feelings between students, the domain of the attribute TYPE is discrete: $D_{TYPE} = \{a, b, c, d, e\}$, where each symbol corresponds to a standard level of feeling between two people; e.g., a corresponds to deep dislike, c to indifference, and e to deep liking.

Having two kinds of domain (continuous and discrete) induces two modes of representation for possibility distributions. The membership functions for the fuzzy sets used as values of the levels in the relation PERSON are shown in Figure 6.2.

Trapezoidal membership functions are usually good enough in practical applications. In fact, slight changes in the form of membership functions (which in any case can hardly be identified with high precision—see Section 6.2.1.5) make little difference to the evaluation of queries. Quadruples can be used for fuzzy values just as well as for precise values: the two first elements of the quadruple bound the set of values with possibility = 1, and the other two define the "spreads" of the distribution on either side of this set. For example, "good" is represented

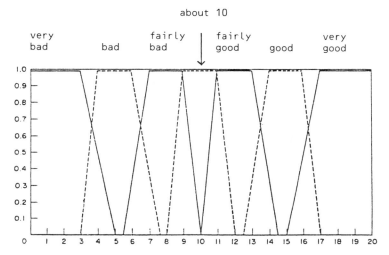

Figure 6.2. Representation of fuzzy marks.

by $(14, 16, 1.5, 1)$, $[10, 12]$ by $(10, 12, 0, 0)$, and 15 by $(15, 15, 0, 0)$. Likewise, "young" and "about-21" are represented by trapezoidal distributions and therefore by quadruples; for instance, "young" is represented by $(18, 23, 2, 2)$, "about-21" by $(20, 22, 1, 1)$, and "about-24" by $(23, 25, 1, 1)$. Finally, the missing value "unknown" is represented by the particular possibility distribution $\mu_1(d) = 1$, $\forall\, d \in D$, and the missing value "inapplicable" by the distribution $\mu_0(d) = 0$, $\forall\, d \in D$.

Moreover, a fuzzy description like "bad-to-very-bad" is interpreted as the convex hull of the union (defined by "max") of the two fuzzy sets representing "bad" and "very bad." Observe that the convex hull is again trapezoidal and so "bad-to-very-bad" is therefore represented by $(0, 6, 0, 1.5)$. When a possibility distribution is not unimodal (e.g., $[10, 12] \cup [15, 18]$)—which does not happen in our example—it is to be split into unimodal parts represented by quadruples (cf. Chapter 2, Section 2.5.1).

On the other hand, for a discrete domain like the attribute TYPE, we represent a possibility distribution as follows: let μ be a possibility distribution defined on a domain $D = \{d_1, \ldots, d_n\}$ such that $\mu(d_i) = a_i$, $\forall\, i$. μ will be represented in memory by the set $\{a_{i_1}/d_{i_1}, \ldots, a_{i_k}/d_{i_k}\}$, where $d_{i_1}, \ldots, d_{i_k}$ are the elements of D on which μ takes a nonzero value. For example, we would represent the value "friendship" of the relation FEELING by $\{0.2/c, 1/d, 1/e\}$ (meaning that there is possibility 0.2 that the sentiment is one of indifference, possibility 1 that it is liking, and possibility 1 that it is deep liking); the value "great-friendship" would be represented by $\{0.8/d, 1/e\}$, and so on.

In this representation, the missing value "unknown" will be represented by $D = \{1/d_1, \ldots, 1/d_n\}$ and the missing value "inapplicable" by $\varnothing$.

6.3.2. Examples of Queries

Now consider the following queries:
"Find all the students

1. whose attainment in math is between 14 and 16 in the first term;
2. whose attainment in math is at least 12 in the second term;
3. whose attainment in math is much-greater-than-10 in the first term;
4. whose attainment in math is at least "good" in the first term;
5. whose attainment in math is definitely better than "good" in the first term;
6. who are at least fairly good in math and at best fairly good in physics in the first term;
7. who are good in math or in physics in the first term;
8. whose attainment in physics is much better in the second term than in the first term;
9. who are friends of Jane;
10. who are friends of a student who is more than 22 years old;
11. who are good at science in the first term."

Note that queries 1–9 involve a single relation while query 10 involves the two relations PERSON and FEELING; the last query is more complex and we shall have to define the relationship between (global) attainment in science and attainments in math and in physics in order to answer it.

Consider first the queries involving one relation. For each query, the response is obtained by applying a Θ-selection to a relation (PERSON or FEELING) followed by a projection. We thus translate

- Query 1 by the projection of $\sigma(\text{PERSON}; \text{M1} = [14, 16])$ onto the attribute NAME.
- Query 5 by the projection of

$$\sigma(\text{PERSON}; \text{M1 "much-greater-than" "good")}$$

onto the attribute NAME.

- Query 8 by the projection of

$$\sigma(\text{PERSON}; \text{P2 "much-greater-than" P1})$$

onto the attribute NAME.
- Query 9 by the projection of

$$\sigma(\text{FEELING}; \text{NAME2} = \text{"Jane"} \wedge \text{TYPE} = \text{"friendship"})$$

onto the attribute NAME1.

Observe that queries 1–5 involve atomic conditions of the form $A \ominus a$, queries 6, 7, and 9 compound conditions of the form $A \ominus a$, and query 8 an atomic condition of the type $A \ominus B$. We also use fuzzy comparators such as "much-greater-than"; this by the way is represented by fuzzy relation

$$\mu_{\text{much-greater-than}}(u, v) = \begin{cases} 0 & \text{if } v - u \leq 2 \\ (v - u)/2 - 1 & \text{if } 2 \leq v - u \leq 4 \\ 1 & \text{if } v - u \geq 4 \end{cases}$$

and shown in Figure 6.3.

The set of elements of a relation (PERSON or FEELING) which possibly or necessarily satisfies a condition $A \ominus a$ has been defined by, respectively, formula (6.15) or formula (6.16) (see 6.2.1.3), and the set of elements possibly or necessarily satisfying a condition $A \ominus B$ by, respectively, formula (6.29) or formula (6.30) (see 6.2.1.4). The result of the $\ominus$-selection for a query can be calculated. The final response, obtained by projection of each of the fuzzy sets making up the result of the $\ominus$-selection, will be denoted by a pair of fuzzy sets $(R - \Pi; R - N)$. For example, the response to query 1 is denoted by $(R_1 - \Pi; R_1 - N)$,

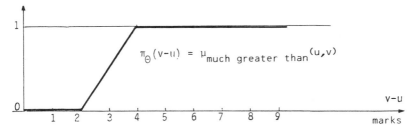

Figure 6.3. Representation of a fuzzy comparator.

where $(R_1 - \Pi)$ represents the set of names of students who possibly answer to the description "their attainment in maths is between 14 and 16 in the first term." Thus we obtain the following responses:

Query 1: $R_1 - \Pi = \{1/\text{Tom}, 0.3/\text{Jane}, 1/\text{Joe}, 1/\text{Jack}\}$
           $R_1 - N = \{1/\text{Tom}, 1/\text{Joe}\}$
Query 2: $R_2 - \Pi = \{1/\text{Tom}, 1/\text{Jane}, 1/\text{Joe}, 1/\text{Jack}\}$
           $R_2 - N = \{1/\text{Tom}, 1/\text{Joe}\}$
Query 3: $R_3 - \Pi = \{1/\text{Tom}, 0.7/\text{Jane}, 1/\text{Joe}, 1/\text{Jack}\}$
           $R_3 - N = \{1/\text{Tom}, 1/\text{Joe}\}$
Query 4: $R_4 - \Pi = \{1/\text{Tom}, 0.6/\text{Jane}, 1/\text{Joe}, 1/\text{Jack}\}$
           $R_4 - N = \{1/\text{Tom}, 1/\text{Joe}\}$
Query 5: $R_5 - \Pi = \{0.1/\text{Tom}, 0.4/\text{Joe}, 1/\text{Jack}\}$
           $R_5 - N = \{0.1/\text{Tom}\}$
Query 6: $R_6 - \Pi = \{1/\text{David}, 1/\text{Bob}, \quad 1/\text{Jane}, \quad 1/\text{Jill}, \quad 1/\text{Jack},$
           $0.3/\text{Joe}\}$
           $R_6 - N = \{0.5/\text{David}, 1/\text{Bob}, 0.5/\text{Jane}, 1/\text{Jill}\}$
Query 7: $R_7 - \Pi = \{1/\text{Tom}, \quad 0.6/\text{David}, \quad 1/\text{Bob}, \quad 1/\text{Jane}, \quad 1/\text{Jill},$
           $1/\text{Joe}, 1/\text{Jack}\}$
           $R_7 - N = \{1/\text{Tom}, 1/\text{Jane}, 0.5/\text{Jill}, 1/\text{Joe}\}$
Query 8: $R_8 - \Pi = \{1/\text{Tom}, 0.6/\text{Bob}, 0.1/\text{Joe}, 0.3/\text{Jack}\}$
           $R_8 - N = \varnothing$
Query 9: $R_9 - \Pi = \{1/\text{Jill}\}$
           $R_9 - N = \{0.8/\text{Jill}\}$

Query 10 involves the two relations PERSON and FEELING and we obtain the response by a Θ-join (see Section 6.2.2). Thus we must calculate the projection of

$$\sigma(\text{PERSON} \times \text{FEELING}; \text{NAME} = \text{NAME2} \wedge \text{TYPE} = \text{"friendship"} \wedge \text{AGE} \geqslant 22)$$

onto the attribute NAME1.

We thus obtain the following response:

$$R_{10} - \Pi = \{0.2/\text{David}, 1/\text{Jane}, 1/\text{Jill}, 1/\text{Jack}\}$$

$$R_{10} - N = \{0.8/\text{Jane}, 1/\text{Jack}\}$$

Finally, consider the last query, which is more complex because it introduces a compound universe of discourse—"science." The relationship between attainment in math and in physics, and attainment in science, can be defined by a relation such as the following:

M (math)	P (physics)	S (science)
Good	Fairly good to good	Good
Fairly good to good	Good	Good
. . .	. . .	. . .
Fairly good to good	About 10	Fairly good
Fairly good	Fairly good	Fairly good
. . .	. . .	. . .

Thus the query "Find all students who are good at science in the first term" can be translated into the query "Find all the students who are good in math and fairly good to good in physics or fairly good to good in math and good in physics in the first term." We obtain the response

$$R_{11} - \Pi = \{1/\text{Tom}, 1/\text{Jane}, 1/\text{Jill}, 1/\text{Jack}\}$$
$$R_{11} - N = \{0.5/\text{Jane}, 0.5/\text{Joe}\}$$

Similarly, the query "Find all the students who are fairly good in science in the second term" has the response

$$R - \Pi = \{0.3/\text{Tom}, 0.6/\text{Bob}, 0.6/\text{Jane}, 1/\text{Jill}\}$$
$$R - N = \{0.5/\text{Jill}\}$$

All the above questions are handled automatically by MACLISP procedures found in the program given in the appendix. These procedures are based on the fuzzy pattern-matching described in Cayrol, Farreny, and Prade [9, 10]. This fuzzy pattern-matching generalizes the pattern-matching systems developed in artificial intelligence (see [40], for instance), in that they use possibility and necessity measures evaluated on [0, 1] to calculate matching measures between two symbolic expressions; they take account of the semantic representation (in terms of possibility distributions) attached to each atom. The possibility and necessity measures are calculated directly from practical representations of attribute values (by a quadruple for a continuous domain, and in the form $\{a_{i_1}/d_{i_1}, \ldots, a_{i_k}/d_{i_k}\}$ when the domain is discrete).

6.4. Conclusion

We have described an extended relational algebra that can deal with incomplete, fuzzy, or uncertain data and that can handle vague queries.

Possibility theory is seen to be an appropriate framework for this problem. The associated query language is implemented in MACLISP; first experience gives results in agreement with intuition: responding to a query in terms of elements that possibly and elements that necessarily (or surely or certainly if one prefers) satisfy its terms seems quite natural when the available data are incomplete. The fact that "possibly" and "necessarily" can be quantified constitutes an extra attraction of an approach that, despite its apparent sophistication, is relatively easy to carry out. In addition, for either complete or partial data, vague queries can be handled.

The query language can be further developed to deal with notions of cardinality, for example. Another interesting problem arising in connection with fuzzy knowledge is that of summarizing the contents of a database, whether these be complete or incomplete (see, for example, [17 and 42]).

Appendix: Computer Program

In this appendix we present the program that handles queries concerning the example in Section 6.3. In the first part we describe the representation of data in the implementation, then we give an account of the practical representation of queries, and finally comment on the procedures for handling these queries. The program is written in MACLISP and implemented on DPS 8 (for an introduction to LISP the reader may consult Winston and Horn [15]).

A.1. Data Structures

The database (DB) is a list of atoms $(R_1, \ldots, R_k)$, each atom representing an extended relation. Atom R_i (e.g., PERSON, FEELING) has two properties: ATT and DAT. The content of the property ATT is the list of attributes of R_i. The content of the property DAT is a list of atoms D_j, each one representing an n-tuple of R_i. Each atom D_j has the property VAL containing the list of values D_j for various attributes; this list is of the form

$$((a_1 A_1) \cdots (a_n A_n))$$

where $A_1, \ldots, A_n$ are the attributes and $a_1, \ldots, a_n$ are labels representing values of D_j over the respective attributes. For example, the values "fairly good" [10, 12], "about 24" are, respectively, denoted by the labels "FG," 10:12, ABO24Y.

The practical representation of the possibility distributions associated with the labels is kept in a list in the following way: for attribute values whose domains are continuous, the list CODUNIV contains pairs (label code), where code is either a quadruple or else the value 'NF when the label represents a precise value. For discrete domains, the list CODUNID keeps the pairs (label code) where here code is itself a list of pairs (element-of-domain degree). Note that the user has the option of defining new labels and giving the associated code at the moment the program uses this code. Let us also specify that the values "inapplicable" (INA) and "unknown" (UNK) are handled separately since the code associated with them will depend on the domain. The comparators used are parametrized in the following way: for a comparator Θ, we can write $\mu_\Theta(u, v) = \pi_\Theta(u - v)$, where π_Θ is a possibility distribution encoded as a quadruple (e.g., "much-greater-than" denoted by "mg" is encoded as $(4 \, N \, 2 \, 0)$, where N is sufficiently large compared with the elements of the domain). The code for a comparator is kept in the property CODREL.

Finally, each attribute has properties NAT and STATE, which, respectively, contain the nature of the associated domain (continuous or discrete) and the state of the domain (simple or compound); these items of information can be provided interactively by the user at a suitable moment, or else entered at the start in the file DATA. The function INIT, for example, fills in the property STATE for each attribute. Here is the part of the file DATA that contains the internal data representation:

```
(setq db '(person feeling))
(putprop 'person '(name m1 m2 p1 p2 age) 'att)
(putprop 'person '(e1 e2 e3 e4 e5 e6 e7) 'dat)
(putprop 'e1 '((tom name) (15 m1) (FG m2) (UNK p1) (14:16 p2)
(young age)) 'val)
(putprop 'e2 '((david name) (FB m1) (G m2) (FG p1) (INA p2)
(20 age)) 'val)
(putprop 'e3 '((bob name) (BVB m1) (10:12 m2) (13:20 p1)
(G p2) (22 age)) 'val)
(putprop 'e4 '((jane name) (FG m1) (VG m2) (14 p1) (10:12 p2)
(AB021Y age)) 'val)
(putprop 'e5 '((jill name) (AB01C m1) (FG m2) (G p1)
(AB012 p2) (young age)) 'val)
(putprop 'e6 '((joe name) (14:16 m1) (VG m2) (G p1) (15 p2)
(AB024Y age)) 'val)
(putprop 'e7 '((jack name) (UNK m1) (B m2) (AB013 p1) (FG p2)
(22:25 age)) 'val)

(setq coduniv '((FG (11 13 1 1.5)) (G (14 16 1.5 1))
(VG (17 20 2 0)) (FB (7 9 1.5 1)) (B (4 6 1 1.5))
(VB (0 3 C 2)) (BVB (1 6 C 1.5)) (ABC10 (9 11 1 1))
(22:25 (22 25 0 0)) (young (18 23 2 2)) (AB021Y (20 22 1 1))
(14:16 (14 16 0 0)) (1C:12 (10 12 0 C)) (13:2C (13 2C 0 0))
(AB012 (11 13 1 1)) (AB013 (12 14 1 1))))

(putprop 'feeling '(name1 name2 type) 'att)
(putprop 'feeling '(s1 s2 s3 s4) 'dat)
(putprop 's1 '((david name1) (jane name2) (slfri type)) 'val)
(putprop 's2 '((jill name1) (jane name2) (fri type)) 'val)
(putprop 's3 '((jack name1) (joe name2) (strfri type)) 'val)
(putprop 's4 '((jane name1) (bob name2) (fri type)) 'val)
```

```
(setq coounid '((slfri ((3 0.3) (4 1) (5 0.5)))
(fri ((3 C.2) (4 1) (5 1))) (strfri ((4 0.8) (5 1)))))
(init 'state 's (appenc (get 'person 'att)
                        (get 'feeling 'att)))
;FILLS IN THE PROPERTY "STATE" FCR EACH ATTRIBUTE
(setq N 1C0) (setq I 2) (setq J 2)
(putprop 'eq '(0 0 0 0) 'codrel)
(putprop 'gr (list 0 N 0 C) 'codrel)
(putprop 'sm (list (minus N) C 0 0) 'codrel)
(putprop 'mg (list 4 N 2 0) 'codrel)
(putprop 'ms (list (minus N) (minus 4) 0 2) 'codrel)
```

A.2. Representation of Queries

In our example, a query can be translated into a selection followed by a projection. The selection yields a set of objects satisfying a certain condition $\mathscr{C}$. In practice, the condition $\mathscr{C}$ is translated as a descriptor P and selection by a (fuzzy) matching of a relation by a descriptor.

For reasons of readability, at the external level we have simplified forms of descriptors; a procedure DESCRIPTOR then transforms each simplified form p_i into an associated descriptor P_i for use in the program. The file DATA contains the simplified descriptors $p_1, \ldots, p_{11}$ associated with the queries $q_1, \ldots, q_{11}$ of the example. In the following we shall comment on the general form of the descriptors used.

We here call an elementary pair a list of two elements of which the second is an atom (an attribute name); an n-tuple of a relation thus has as its property VAL a list of elementary pairs of which the first element is a label.

A descriptor P is of the form (AND $P_1, \ldots, P_k$) or (OR $P_1, \ldots, P_k$) or P_k, where $P_1, \ldots, P_k$ are simple descriptors. A simple descriptor is a list of elementary pairs of which the first term is

- An expression: atom (label), capturing variable (e.g., $>X$) or linked variable (e.g., $<X$); or
- A pattern-matching function of the form (APPLY ⟨procedure⟩) where the procedure has no capturing variable.

and whose second term is an atom (a possibly compound attribute name, e.g., SCIENCE). The forms (AND $P_1, \ldots, P_k$), (OR $P_1, \ldots, P_k$) are used when we wish to apply a logical combination of descriptors to a datum (e.g., P_7).

The principle of pattern-matching is to compare each elementary pair of the descriptor P with the associated pair (i.e., with the same second term) of an element D of the relation. This comparison gives two indices (possibility and necessity) for measuring the compatibility between the contents of the pairs; global measures (possibility and necessity that D is compatible with P) are then obtained by aggregating the intermediate indices by the operator min.

The comparison between the pairs $(>X\ A)$ and $(b\ A)$ (belonging, respectively, to P and D) gives indices equal to 1 and assigns the value b to the variable X; in the case $(<X\ A)$ compared with $(a\ A)$, X is instantiated by the previously captured value b and $(b\ A)$ is compared with $(a\ A)$.

In the case of a pattern-matching function, the procedure is instantiated before applying it to the elementary pair associated with the domain D. The pattern-matching functions are used for expressing atomic conditions of the form $A \Theta B$ or $A \Theta a$, where Θ is a comparator other than equality. Thus the procedures determine the kind of processing to be applied to the values in order to obtain the indices corresponding to the various conditions (see Sections 6.2.1.3 and 6.2.1.4).

Simplified descriptors differ from descriptors only in the expression of their possible pattern-matching functions: the object of the function DESCRIPTOR is to construct the procedure for the pair (APPLY ⟨procedure⟩) from the contents of the associated simplified descriptor.

Here is the extract from the file DATA that contains the descriptors:

```
(setq p1 '((14:16 m1)))
(setq p2 '((12: m1)))
(setq p3 '(((toapply process 'mg '(1C m1)) m1)))
(setq p4 '(((toapply process 'gr '(G m1)) m1)))
(setq p5 '(((toapply process 'mg '(G m1)) m1)))
(setq p6 '(((toapply process 'sm '(FC m1)) m1)
((toapply process 'gr '(FG p1)) p1)))
(setq p7 '(or ((G m1)) ((G p1))))
(setq p8 '((>%x p1) ((toapply procsym 'mg '(<%x p1)) p2)))
(setq p9 '((jane name2) (fri type)))
(setq p1C '((>x name) ((toapply process 'gr '(22 age)) age)
(<x name2) (fri type)))
(setq p11 '((G sciences)))
```

A.3. Description of Implemented Procedures

A query bearing on a relation R is translated into a simplified descriptor p and the associated selection is obtained by calling the function SELEC. Note that if the query bears on two relations, a join must be made; we give later an appropriate procedure that is faster than selection on a Cartesian product.

```
(defun descriptor (p)
  (cond ((null p) nil)
        ((eq (car p) 'anc)
         (cons 'and (mapcar (function descriptor) (cdr p))))
        ((eq (car p) 'or)
         (cons 'or (mapcar (function descriptor) (cdr p))))
        ((pair p)
         (cond ((eq (caar p) 'toapply)
                (list (translation (car p)) (cadr p)))
               (t p)))
        (t (cons (descriptor (car p))
                 (descriptor (cdr p))))))
;TRANSFORMS THE SIMPLIFIED FORM "P" INTO
;A DESCRIPTOR
```

```
(defun translation (l)
   (list (car l)
         (list 'lambda '(z)
               (list (cadr l) (cacdr l) 'z (cadddr l)))))
(defun pair (l)
   (and (eq (length l) 2) (atom (car (last l)))))
;DEFINES AN ELEMENTARY PAIR

(defun min2 (l1 l2)
   (list (min (car l1) (car l2))
         (min (cadr l1) (cadr l2))))
(defun max2 (l1 l2)
   (list (max (car l1) (car l2))
         (max (cadr l1) (cadr l2))))
;"L1" AND "L2" ARE PAIRS CF DEGREES FROM [0,1]
;MIN2 (MAX2) BUILDS THE PAIR CONTAINING THE MIN (MAX)
;OF THE FIRST TERMS AND THE MIN (MAX) OF THE SECCND TERMS

(setq bag nil)

(defun selec (p r) (selec1 (descriptor p) (get r 'dat)))
;"P" IS A DESCRIPTOR UNDER SIMPLIFIED FORM
(defun selec1 (p l)
   (let ((-lp nil) (-ln nil))
      (mapcar (function (lambda (x) (let ((-f (patmat p x)))
               (cond ((not (zerop (car -f)))
                      (setq -lp (append -lp
                      (list (list x (car -f)))))))
               (cond ((not (zerop (cadr -f)))
                      (setc -ln (append -ln
                      (list (list x (cadr -f)))))))))) l)
      (list -lp -ln)))
;"R" IS AN ATOM REPRESENTING A RELATICN
;"L" IS THE LIST OF THE ATOMS REPRESENTING THE DATA FROM "R"
;"P" IS A DESCRIPTOR
```

(SELEC P R)

- Calls the function DESCRIPTOR, which transforms p into a descriptor P.
- Successively processes each datum D of R (PATMAT P D).
- Returns a list made up of two lists of evaluated data:

$$(((Di_1 \, di_1) \cdots (Di_k \, di_k))((Dj_1 \, dj_1) \cdots (Dj_p \, dj_p)))$$

[Represents $\sigma\Pi(R; C)$]   [Represents $\sigma N(R; C)$]

```
(defun patmat (p d) (patmat1 p (get c 'val)))
;"D" IS AN ATOM REPRESENTING AN ELEMENT OF THE RELATION
(defun patmat1 (p d)
   (cond ((null p) '(1 1))
         ((eq (car p) 'anc) (and (cdr p) d))
         ((eq (car p) 'cr) (or (ccr p) d))
         ((pair p) (patmat2 p d))
         (t (min2 (patmat1 (car p) d) (patmat1 (cdr p) d))))))
;"D" IS THE LIST OF THE VALUES OF A DATUM
```

(PATMAT P D)

- Processes the contents of the property VAL of datum D and the descriptor P.
- Decomposes P into elementary descriptors.

- Successively processes each elementary pair of *P* by PATMAT2 and combines the successive results by min.

```
(defun patmat2 (p d)
  (cond ((null d) '(0 C))
        ((compose (cadr p)) (patcompo p d))
        ((equal (cadr p) (cadr (car d))) (compare p (car d)))
        (t (patmat2 p (cdr d)))))
;"P" IS AN ELEMENTARY PAIR

(defun compose (a)
  (cond ((equal (get a 'state) 's) nil)
        ((equal (get a 'state) 'c) t)
        (t (print a)
           (prin1 "compound universe yes-no")
           (cond ((eq (atomcar (read)) 'n)
                  (putprop a 's 'state) nil)
                 (t (putprop a 'c 'state))))))
;CHECKS IF THE UNIVERSE CF ATTRIBUTE "A" IS COMPOUND
(defun patcompo (p d)
  (cond ((get (cadr p) (car p))
         (patmat1 (get (cadr p) (car p)) d))
        (t (print 'define) (prin1 p)
           (putprop (caor p) (read) (car p))
           (patmat1 (get (cadr p) (car p)) d))))
;THE DEFINITION CF A VALUE ON A COMPOUND UNIVERSE IS STORED
;IN A PROPERTY ATTACHED TO THAT UNIVERSE
;THE DEFINITION OF "P" IS GIVEN UNDER THE FORM (OR P1 P2 ... Pk)
```

(PATMAT2 P D)

- Tests whether the attribute of the elementary pair *P* is compound and if so calls the procedure PATCOMPO.
- Seeks in *D* the elementary pair associated with *P* and, if it exists, calls the procedure COMPARE.

```
(defun compare (p d)
  (cond ((and (atom (car p)) (eq (atomcar (car p)) '>))
         (setq bag (cors (cons (atomcdr (car p)) (car d)) bag))
         '(1 1))
        ((and (atom (car p)) (eq (atomcar (car p)) '<))
         (egalf (list (cdr (assq (atomcdr (car p)) bag))
                      (cadr p)) d))
        ((atom (car p)) (egalf p d))
        ((eq (caar p) 'toapply)
         (map1 (instanc (cdar p)) d))))
;"P AND "D" ARE  ELEMENTARY PAIRS

(defun and (l d)
  (cond ((null l) '(1 1))
        (t (min2 (patmat1 (car l) d) (and (cdr l) d)))))
(defun or (l d)
  (cond ((null l) '(0 C))
        (t (max2 (patmat1 (car l) d) (or (cdr l) d)))))
;DATUM "D" IS MATCHED AGAINST A COMBINATION CF PATTERNS

(defun map1 (l x)
  (cond ((null l) '(1 1))
        (t (apply (car l) (list x)))))

(defun instanc (l)
  (cond ((null l) nil)
        ((and (atom l) (eq (atomcar l) '<))
         (cdr (assq (atomcdr l) bag)))
        ((atom l) l)
        (t (cons (instanc (car l)) (instanc (cdr l)))))))
;INSTANTIATION FUNCTION
```

(COMPARE P D)

- Takes as arguments two elementary pairs having the same second term.
- Returns a list of two degrees in $[0, 1]$ representing the possibility and the necessity that the contents of P and D are compatible.
- Calls the comparison procedures appropriate to the first term of P: if this is a linked variable, the comparison between D and the instantiation of P is carried out using EGALF; if it is an atom, EGALF is also called; if it is a matching function, the procedure it contains is instantiated by INSTANC and this procedure is applied to D by MAP1.

The instantiation mechanism is based on an A-list BAG, which holds the different instantiations of the variables.

```
(defun jcin (p r1 r2)

  (joint (descriptor p) (get r1 'dat) (get r2 'dat)))
;THE RELATIONS "R1" AND "R2" ARE JOINED UNDER THE CONDITION
;EXPRESSED BY "P"
(defun jcint (p l1 l2)
  (let ((-lp nil) (-ln nil))
    (mapcar (function (lambda (x)
      (mapcar (function (lambda (y)
        (let ((-f (patmat1 p (append (get x 'val) (get y 'val)))))
          (cond ((not (and (zerop (car -f))
                           (zerop (cadr -f))))
                 (let ((-p (prod x y)))
                   (cond ((not (zerop (car -f)))
    (setc -lp (append -lp (list (list -p (car -f))))))
                         (cond ((not (zerop (cadr -f)))
    (setc -ln (append -ln (list (list -p (cadr -f)))))
    ))))))))) l2))) l1)
    (list -lp -ln)))

(defun prod (d1 d2)
  (prodbis d1 d2 (implode (append (explode d1) (explode d2)))))
(defun prodbis (d1 d2 d)
  (putprop d (append (get d1 'val) (get d2 'val)) 'val)
  d)
```

$(\text{JOIN } p \ R_1 \ R_2)$

- Transforms p into a descriptor P.
- Carries out a direct join on R_1 and R_2 without calculating their Cartesian product.
- Works with a descriptor P and the lists of data L_1 and L_2 of the relations R_1 and R_2 using $(\text{JOINT } P \ L_1 \ L_2)$.

Algorithm:

- Traverse the list L_1: $\forall \ d_1 \in L_1$.
- Traverse the list L_2: $\forall \ d_2 \in L_2$.

- Calculate the associated degrees by comparing the contents of *P* with those of d_1 and d_2.
- If one of these degrees is non-null, form the datum produced by PROD.

The following procedures are used to project a selection or a join onto an attribute:

```
(defun projec (ll a)
  (list (reduc (projec1 (car ll) a))
        (reduc (projec1 (cadr ll) a))))
;"LL" IS A LIST OF TWO LISTS OF GRADED DATA
;"PROJEC" REPLACES THE DATA BY THEIR VALUES ON "A"

(defun projec1 (l a)
  (cond ((null l) nil)
        (t (cons (list (ass a (get (caar l) 'val))
                       (cadar l))
                 (projec1 (cdr l) a)))))
;"PROJEC1" PROCESSES ATOMIC DATA

(defun ass (a l)
  (cond ((null l) nil)
        ((equal a (cadar l)) (caar l))
        (t (ass a (cdr l)))))
;LOOKS FOR THE ATOM OF THE PAIR HAVING "A" AS SECOND ELEMENT
;IN THE LIST "L" OF ELEMENTARY PAIRS

(defun reduc (l)
  (cond ((null l) nil)
        (t (reducbis l (assq (caar l) (cdr l))))))
(defun reducbis (l a )
  (cond ((null a) (cons (car l) (reduc (cdr l))))
        (t (reduc (cons (list (caar l)
                              (max (cadar l) (cadr a)))
                        (supp a (cdr l)))))))

(defun supp (c l)
  (cond ((null l) nil)
        ((equal (car l) c) (cdr l))
        (t (cons (car l) (supp c (cdr l))))))
;"L" IS A LIST OF PAIRS, "C" IS A PAIR , "SUPP"
;YIELDS THE LIST OBTAINED FROM "L" BY REMOVING THE
;FIRST OCCURRENCE OF "C"
```

(PROJECT LL A)

- Takes as arguments a list of two lists of graded data which represent the result of a selection, and an attribute name.
- Successively processes each of the lists making up LL by the functions PROJEC1 and REDUC.

REDUC is applied to a list of pairs (atom degree) and combines two pairs having the same atom by taking the max of the degrees.

The following procedures concern the actual processing of fuzzy values:

```
;"X1" AND "X2" ARE ELEMENTARY PAIRS HAVING THE SAME
;SECOND TERM AND THE FIRST TERM CF WHICH IS A LABEL

(defun egalf (x1 x2)
  (cond ((eq (car x2) 'UNK) '(1 C))
        ((eq (car x2) 'INA) '(0 C))
        ((discrete (cadr x1)) (egalf2 x1 x2))
        (t (egalf1 x1 x2))))))

(defun discrete (u)
  (cond ((equal (get u 'nat) 'c) nil)
        ((equal (get u 'nat) 'd) t)
        (t (print u) (prin1 "discrete universe yes-no")
           (cond ((eq (atomcar (read)) 'n)
                  (putprop u 'c 'nat) nil)
                 (t (putprop u 'c 'nat) t)))))
;"DISCRETE" YIELDS "T" IF UNIVERSE "U" IS DISCRETE
(defun egalf1 (x1 x2)
  (let ((-c1 (code (car x1))) (-c2 (code (car x2))))
       (cond ((eq -c1 'nf)
              (setq -c1 (list (car x1) (car x1) 0 C))))
       (cond ((eq -c2 'nf)
              (setq -c2 (list (car x2) (car x2) 0 C))))
       (list (poss1 -c1 -c2) (nec1 -c1 -c2))))

(defun code (x)
  (cond ((null (assq x coduniv))
         (print "give the code") (prin1 x)
         (print "nf in case of a non fuzzy atom")
         (setq coduniv (cons (list x (read)) coduniv))))
  (cadr (assq x coduniv)))
;LOOKS FCR THE CODE ATTACHED TO "X" IN THE CCNTINUOUS CASE

(defun egalf2 (x1 x2)
  (let ((-c1 (codd (car x1))) (-c2 (codd (car x2))))
       (cond ((eq -c1 'nf) (setq -c1 (list (list (car x1) 1)))))
       (cond ((eq -c2 'nf) (setq -c2 (list (list (car x2) 1)))))
       (list (poss2 -c1 -c2) (nec2 -c1 -c2))))

(defun codd (x)
  (cond ((null (assq x codunid))
         (print "give the code") (prin1 x)
         (print "nf in case of a precise value")
         (setq codunid (cons (list x (eval (read))) codunid))))
  (cadr (assq x codunid)))
;LOOKS FCR THE CODE ATTACHED TO "X" IN THE DISCRETE CASE
```

(EGALF XI X2)

- Calculates $\Pi(a_1 \mid a_2)$ and $N(a_1 \mid a_2)$ where a_1, or a_2, is the label of
  $X1$, or $X2$.
- Calls for this purpose the function EGALF1 if the universe associated with $X1$ and $X2$ is continuous, EGALF2 otherwise.

The above calculations are carried out on the codes attached to the labels by the functions POSS1, NEC1 in the continuous case, and POSS2, NEC2 in the discrete case.

```
(defun process (z x1 x2)
  (cond ((eq (car x1) 'UNK) '(1 0))
        ((eq (car x1) 'INA) '(0 0))
        (t (let ((-c1 (ccce (car x1))) (-c2 (code (car x2))))
        (cond ((eq -c1 'nf) (setq -c1 (list (car x1) (car x1) 0 0))))
        (cond ((eq -c2 'nf) (setq -c2 (list (car x2) (car x2) 0 0))))
        (list (poss1 (sumf (encode z) -c2) -c1)
              (nec1 (sumf (encode z) -c2) -c1))))))
;"Z" DENOTES A COMPARATOR

(defun encode (z)
  (cond ((null (get z 'codrel))
         (print "give the code") (prin1 z)
         (putprop z (read) 'codrel)))
  (get z 'codrel))
;LOOKS FOR THE CODE ATTACHED TO "Z"

(defun sumf (t1 t2)
  (mapcar (function plus) t1 t2))
;ADDS TWO QUADRUPLES TERM BY TERM
```

(PROCESS Z X1 X2)

- Calculates $\Pi(a_2 \circ \Theta \mid a_1)$ and $N(a_2 \circ \Theta \mid a_1)$, where Θ is the comparator designated by Z.
- Uses the codes attached to Z, a_1, a_2 and the functions POSS1 and NEC1.
- Uses SOMMEF to calculate the code attached to $a_2 \circ \Theta$.

```
(defun procsym (z x1 x2)
  (cond ((or (eq (car x2) 'INA) (eq (car x1) 'INA)) '(0 0))
        ((or (eq (car x2) 'UNK) (eq (car x1) 'UNK)) '(1 0))
        (t (let ((-c1 (code (car x1)))
                 (-c2 (code (car x2))))
        (cond ((eq -c1 'nf)
               (setq -c1 (list (car x1) (car x1) 0 0))))
        (cond ((eq -c2 'nf)
               (setq -c2 (list (car x2) (car x2) 0 0))))
        (list (poss1 (sumf (encode z) -c2) -c1)
              (neces z -c1 -c2)))))))
;"Z" IS AN ATOM REPRESENTING A COMPARATOR
;"PROCSYM" PROCESSES THE DATA SYMMETRICALLY
```

(PROCSYM Z X1 X2)

- Calculates $\Pi(\Theta \mid (a_1, a_2))(= \Pi(a_2 \circ \Theta \mid a_1))$ by POSS1.
- Calculates $N(\Theta \mid (a_1, a_2))$ using NECES.

```
(defun neces (z c1 c2)
  (necesbis z c1 c2 (compl z)))
(defun necesbis (z c1 c2 c)
  (minus (sub1 (max
         (poss1 (sumf (car c) c2) c1)
         (cond ((cdr c) (poss1 (sumf (cadr c) c2) c1))
               (t 0)))))))

;"C" IS A LIST CONTAINING ONE OR TWO QUADRUPLES
;FUNCTION "COMPL" YIELDS THE QUADRUPLE(S) ASSOCIATED
```

```
;WITH THE COMPLEMENTARY CF ITS ARGUMENT
(defun ccmpl (z)
  (cond ((eq z 'eq) (list (list C N C 0)
                          (list (minus N) C 0 C)))
        ((eq z 'gr) (list (list (minus N) 0 C 0)))
        ((eq z 'sm) (list (list C N C 0)))
        ((eq z 'mg) (list (list (minus N) 2 C 2)))
        ((eq z 'ms) (list (list (minus 2) N 2 0)))))
```

(NECES Z X1 X2)

- Calculates $N(\Theta \mid (a_1, a_2))$ by $1 - \Pi(a_2 \circ \bar{\Theta} \mid a_1)$.
- Uses the function COMPL to obtain the code attached to the comparator $\bar{\Theta}$ by $\mu_{\bar{\Theta}}(u, v) = 1 - \mu_{\Theta}(u, v)$.

Note that in certain cases, $\bar{\Theta}$ cannot be parametrized by a quadruple but can be represented by a union of two trapezoidal distributions. The function COMPL returns one of these two quadruples for the parameter $\bar{\Theta}$.

```
(defun poss1 (c1 c2)
  (poss1bis c1 c2 (difference (cadr c2) (car c1))
                  (difference (cadr c1) (car c2))))
(defun poss1bis (c1 c2 x y)
  (cond ((ge (plus (difference (cadr c1) (car c1))
                   (difference (cadr c2) (car c2)))
             (max x y)) 1)
        ((greaterp x y)
         (cond ((and (zerop (caddr c2)) (zerop (cacddr c1))) 0)
               (t (born (cuotient (flcat (oifference (car c2)
                                                     (cadr c1)))
                                  (float (plus (cadcdr c1) (caccr c2))))))))
        ((greaterp y x)
         (cond ((and (zerop (caddr c1)) (zerop (cadccr c2))) 0)
               (t (born (quotient
                         (float (oifference (car c1) (cadr c2)))
                         (float (plus (cacddr c2) (caccr c1))))))))))

(defun bcrn (x) (max 0 (minus (sub1 x))))

(defun nec1 (c1 c2)
  (cond ((and (zerop (cacor c1)) (zerop (cadcr c2))
              (zerop (caccdr c1)) (zerop (caccdr c2)))
         (cond ((lessp (car c2) (car c1)) 0)
               ((ge (cadr c1) (cadr c2)) 1)
               (t 0)))
        ((and (zerop (caccdr c1)) (zerop (caccdr c2)))
         (cond ((lessp (cadr c1) (cacr c2)) C)
               (t (min 1
                       (max C (quotient (float (plus (caccr c1) (car c2)
                                                     (minus (car c1))))
                                        (float (plus (cacdr c2) (cadcr c1)))))))))
        ((and (zerop (cacdr c1)) (zerop (cadcr c2)))
         (cond ((lessp (car c2) (car c1)) 0)
               (t (min 1
                       (max C (quotient (float (plus (caccdr c1) (cadr c1)
                                                     (minus (cacr c2))))
                                        (float (plus (cadcdr c2) (cadcdr c1)))))))))
        (t (min 1 (min
                   (max 0 (quotient (float (plus (caddr c1) (car c2)
                                                 (minus (car c1))))
                                    (float (plus (caddr c2) (cacdr c1))))))))))
```

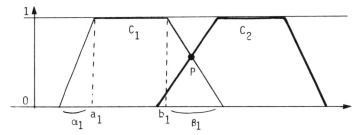

Figure 6.4. Computation of a possibility degree.

```
(max 0 (quotient (float (plus (cadddr c1) (cadr c1)
                               (minus (cadr c2))))
               (float (plus (cacddr c2) (caccdr c1))))))))))))
;"POSS1" ("NEC1")  CALCULATES THE POSSIBILITY (NECESSITY) OF
;"C1" GIVEN "C2"
;"C1" AND "C2" ARE QUADRUPLES
```

The functions POSS1 and NEC1 have quadruples for arguments. Let $C_1 = (a_1\, b_1\, \alpha_1\, \beta_1)$ and $C_2 = (a_2\, b_2\, \alpha_2\, \beta_2)$.

In the case shown in Figure 6.4. (POSS1 C1 C2) returns the ordinate of the point P.

In the case shown in Figure 6.5. (NEC1 C1 C2) returns the ordinate of the point N.

```
(defun poss2 (c1 c2)
  (cond ((null (cdr c2))
         (min (cadar c2) (cond ((assc (caar c2) c1)
                                (cadr (assq (caar c2) c1)))
                               (t 0))))
        (t (max (poss2 c1 (list (car c2)))
                (poss2 c1 (cdr c2)))))))

(defun nec2 (c1 c2)
  (cond ((null (cdr c2))
         (max (minus (sub1 (cadar c2)))
              (cond ((assq (caar c2) c1)
                     (cacr (assq (caar c2) c1)))
                    (t 0))))
        (t (min (nec2 c1 (list (car c2)))
                (nec2 c1 (cdr c2)))))))

;"POSS2" AND "NEC2" TAKE LISTS OF PAIRS (ELEMENT_OF_DOMAIN
; DEGREE ) AS ARGUMENTS
```

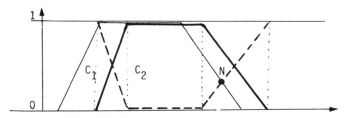

Figure 6.5. Computation of a necessity degree.

The functions POSS2 and NEC2 have as arguments lists of pairs. Let

$$C_1 = ((a_{i1} u_{i1}) \cdots (a_{ik} u_{ik}))$$
$$C_2 = ((a_{j1} u_{j1}) \cdots (a_{j1} u_{j1}))$$

(POSS2 C1 C2) calculates $\sup_{k/v_k > 0} \min(u_k, v_k)$.

(NEC2 C1 C2) calculates $\inf_{k/v_k > 0} \max(u_k, 1 - v_k)$.

To close this appendix we give the details of the processing of a query, Q5 for instance.

Statement. Find all students whose achievement in math is much better than "good" in the first term.

Translation into operations of relational algebra: Projection of σ(PERSON; M_1 "much-greater-than" "good") onto the attribute NAME.

Descriptor p_5 associated with condition C_5 of the selection: C_5 is of the form $A \ominus a$ and for each n-tuple x of the relation PERSON we must calculate Π('good' $\circ$ 'mg' $| M_1(x)$) and N('good' $\circ$ 'mg' $| M_1(x)$), where 'mg' denotes the comparator "much-greater-than." Thus we must use the procedure PROCESS for the calculations, whence we have the form of p_5 which contains only an elementary pair C_5:

$$((\text{apply process 'mg } Z' (G M_1)) M_1)$$

The call to DESCRIPTOR will transform C_5 into a further elementary pair C_5':

$$((\text{apply}(\text{lambda}(Z)(\text{process 'mg } Z'(G M_1)))) M_1)$$

Matching data $e_1, \ldots, e_7$ by P_5: The function PATMAT2 seeks in the value of e_1 the elementary pair associated with P_5, i.e., $(15 M_1)$ and calls the function (COMPARE C_5' '(15 M_1)). The matching function of C_5' then applies the procedure it contains with the parameter Z equal to $(15 M_1)$. The processing proceeds by seeking the codes:

$15 \rightarrow$ 'NF (exact value) translated by (15 15 0 0)

$G \rightarrow$ (14 16 15 1)
'mg $\rightarrow$ (4 N 2 0) with $N = 100$

and calls

(POSS1 '(14 + 14 16 + N 15 + 2 1 + 0) '(15 15 0 0))

(NEC1 '(18 16 + N 3.5 1) '(15 15 0 0))

whence the result (0.1 0.1).

The following functions, which exist in certain versions of LISP, have been redefined here and are stored in a file which is always open during a session:

```
(defun listp (x)
  (eq (typep x) 'list))
;LIST DEFINITION
(setq base 10.)
(setq ibase 10.)
(defun ge (n1 n2)
  (not (lessp n1 n2)))
;DEFINITION OF THE RELATION "GREATER OR EQUAL"
(defun atomcar (x)
  (car (explode x)))
(defun atomcdr (x)
        (implode (cdr (explode x))))

(defun irit (pte va l)
  (mapcar (function (lambda (a)
    (putprop a va pte)))
    l))
;FILLS IN THE PROPERTY "PTE" WITH VALUE "VA" FOR EACH
;ELEMENT IN LIST "L"
```

References

1. ADIBA, M., and DELOBEL, C. (1982). *Bases de Données et Systèmes Relationnels.* Dunod, Paris.
2. BALDWIN, J. F. (1983). A fuzzy relational inference language for expert systems. *Proc. 13th IEEE Int. symp. on Multiple-Valued Logic,* Kyoto, Japan, pp. 416–423.
3. BALDWIN, J. F. (1983). Knowledge engineering using a fuzzy relational inference language. *Proc. IFAC Symposium on Fuzzy Information, Knowledge Representation and Decision Processes,* Marseille, July 19–21, pp. 15–20.
4. BISKUP, J. (1980). A formal approach to null values in database relations. Workshop: *Formal Bases for Databases.* December, 12–14, 1979, CERT-DERI. Toulouse (H. Gallaire and J. M. Nicolas, eds.). Plenum Press, New York.
5. BOSSU, G., and SIEGEL, P. (1985). Saturation, nonomonotonic reasoning, and the closed world assumption. *Artif. Intell.* **25,** 13–63.
6. BUCKLES, B. P., and PETRY, F. E. (1982). A fuzzy representation of data for relational databases. *Fuzzy Sets Syst.,* **7,** 213–226.
7. BUCKLES, B. P., and PETRY, F. E. (1982). Fuzzy databases and their applications. In *Fuzzy Information and Decision Processes* (M. M. Gupta and E. Sanchez, eds.). North-Holland, Amsterdam, pp. 361–371.
8. BUCKLES, B. P., and PETRY, F. E. (1983). Extension of the fuzzy database with fuzzy arithmetic. *Proc. IFAC Symposium, Fuzzy Information, Knowledge Representation and Decision Processes,* Marseille, July 19–21, pp. 409–414.
9. CAYROL, M., FARRENY, H., and PRADE, H. (1980). Possibility and necessity in a pattern matching process. *Proc. IXth. Int. Cong. on Cybernetics,* Namur, Belgium, September 8–13, pp. 53–65.
10. CAYROL, M., FARRENY, H., and PRADE, H., (1982). Fuzzy pattern matching. *Kybernetes,* **11,** 103–116.
11. CODD, E. F. (1979). Extending the database relational model to capture more meaning *ACM Trans. Database Syst.* **4**(4), 397–434.

12. DATE, D. J. (1977). *An Introduction to Data Base Systems.* Addison-Wesley, Reading, Massachusetts.
13. DUBOIS, D., and PRADE, H. (1983). Twofold fuzzy sets: An approach to the representation of sets with fuzzy boundaries based on possibility and necessity measures. *Fuzzy Math. (Huazhong, China)*, 3(4), 53–76.
14. DUBOIS, D., and PRADE, H. (1985). Fuzzy cardinality and the modeling of imprecise quantification. *Fuzzy Sets Syst.*, 16, 199–230.
15. WINSTON, P. H. and HORN, B. K. P. (1981). *LISP.* Addison-Wesley, Reading, Massachusetts.
16. FRESKA, C. (1980). L-FUZZY—An A.I. language with linguistic modification of patterns. AISB Conf. Amsterdam. (Also UCB/ERL M80/10, Univ. of California, Berkeley.)
17. GELENBE, E. (1983). Incomplete representations of information in data bases. Research, Report, No 9, ISEM, Univ. Paris-Sud.
18. GRANT, J. (1979). Null values in a relational data base. *Inf. Process. Lett.*, 6(5), 156–157.
19. GRANT, J. (1979). Partial values in a tabular database. *Inf. Process. Lett.*, 9(2), 97–99.
20. HAAR, R. L. (1977). A fuzzy relational data base system. University of Maryland. Computer Center. TR-586, September.
21. LE FAIVRE, R. (1974). The representation of fuzzy knowledge. *J. Cybernet.*, 4(2), 57–66.
22. LIPSKI, W., Jr. (1979). On semantic issues connected with incomplete information data bases. *ACM Trans. Database Syst.*, 4(3), 262–296.
23. LIPSKI, W., Jr. (1981). On databases with incomplete information. *J. Assoc. Comput. Machinery*, 28(1), 41–70.
24. MONTGOMERY, C. A., and RUSPINI, E. H. (1981). The active information system: A data driven system for the analysis of imprecise data. *Proc. VIIth. Int. Conf. on Very Large Databases,* Cannes, September.
25. NARIN'YANI, A. S. (1980). Sub-definite set—New data-type for knowledge representation. (in Russian). Memo No 4-232 Computer Center. Novosibirsk. URSS.
26. PHILIPS, R. J., BEAUMONT, M. J., and RICHARDSON, D. (1979). AESOP. An Architectural relational database. *Comput. Aided Des.*, 11(4), 217–226.
27. PRADE, H. The connection between Lipski's approach to incomplete information data bases and Zadeh's possibility theory. *Proc. Int. Conf. Systems Methodology,* Washington, D.C., January, 5–9, pp. 402–408.
28. PRADE, H. (1982). Possibility sets, fuzzy sets and their relation to Łukasiewicz logic. *Proc. 12th. Symp. on Multiple-Valued Logic,* Paris, May 24–27, pp. 223–227.
29. PRADE, H. (1982). Modèles mathématiques de l'imprécis et de l'incertain en vue d'applications au raisonnement naturel (358 p.). Thèse d'Etat, Univ. Paul Sabatier, Toulouse.
30. PRADE, H. (1983). Représentation d'informations incomplètes dans une base de données à l'aide de la théorie des possibilités. *Proc. Convention Informatique Latine 83,* Barcelona, Spain, June 6–9, pp. 378–392.
31. PRADE, H. (1984). Lipski's approach to incomplete information databases restated and generalized in the setting of Zadeh's possibility theory. *Inf. Syst.* 9(1), 27–42.
32. PRADE, H. (1983). Do we need a precise definition of membership functions? *BUSEFAL,* No. 14, LSI, University Paul Sabatier, Toulouse, p. 127.
33. PRADE, H., and TESTEMALE, C. (1984). Generalizing database relational algebra for the treatment of incomplete/uncertain information and vague queries. *Inf. Sci.,* 34, 115–143.
34. PRADE, H., and TESTEMALE, C. (1987). Representation of soft constraints and fuzzy

attribute values by means of possibility distributions in databases. In *The Analysis of Fuzzy Information*, Volume 2: *Artificial Intelligence and Decision Systems* (J. Bezdek, ed.), CRC Press, Boca Raton, Florida, pp. 213–229.

35. RUSPINI, E. (1982). Possibilistic data structures for the representation of uncertainty. In *Approximate Reasoning in Decision Analysis* (M. M. Gupta and E. Sanchez, eds.), North-Holland, Amsterdam, pp. 411–415.

36. SIKLOSSY, L., and LAURIERE, J. L. (1982). Removing restrictions in the relational database model: An application of problem-solving techniques. *Proc. National Conf. in Artificial Intelligence* Pittsburg, August.

37. TAHANI, V. (1977). A conceptual framework for fuzzy query processing —A step toward very intelligent database systems. *Inf. Process. Manage.*, **13**, 289–303.

38. UMANO, M. (1982). FREEDOM-0: A fuzzy database system. In *Fuzzy Information and Decision Processes* (M. M. Gupta and E. Sanchez, eds.), North-Holland, Amsterdam, pp. 339–349.

39. UMANO, M. (1983). Retrieval from fuzzy data base by fuzzy relational algebra. *Proc. IFAC Symposium, Fuzzy Information, Knowledge Representation and Decision Processes*, Marseille, July 19–21, pp. 1–6.

40. WINSTON, P. H. (1977). *Artificial Intelligence*. Addison-Wesley, Reading, Massachusetts.

41. WONG, E. A. (1982). Statistical approach to incomplete information in database systems. *ACM Trans. Database Syst.*, **7**(3), 470–488.

42. YAGER, R. (1982). A new approach to the summarization of data. *Inf. Sci.*, **28**, 69–86.

43. ZADEH, L. A. (1978). Fuzzy sets as a basis for a theory of possibility. *Fuzzy Sets Syst.*, **1**(1), 3–28.

44. ZADEH, L. A. (1978). PRUF: A meaning representation language for natural languages. *Int. J. Man-Machine Stud.*, **10**, 395–460.

45. ZADEH, L. A. (1981). Test-score semantics for natural languages and meaning representation via PRUF. SRI International Technical Note No. 247, May 1981, Menlo Park, California. Also in *Empirical Semantics*, Vol. 1 (B. B. Rieger, ed.). Brockmeyer, Bochum, pp. 281–349.

46. PRADE, H., and TESTEMALE, C. (1987). Fuzzy relational data bases: Representational issues and reduction using similarity measures. *J. Am. Soc. Inf. Sci.* **38**(2), 118–126.

47. KUNII, T. L. (1976) DATAPLAN: an interface generator for database semantics. *Inf. Sci.*, **10**, 279–298

48. ZEMANKOVA, M. and KANDEL, A. (1984). *Fuzzy Relational Data Bases: A Key to Expert Systems*. Verlag TÜV Rheinland, Köln.

49. KACPRZYK, J., and ZIOLKOWSKI, A. (1986). Retrieval from data bases using queries with fuzzy linguistic quantifiers. In *Fuzzy Logic in Knowledge Engineering* (H. Prade and C. V. Negoita, eds.), Verlag TÜV Rheinland Köln, pp. 46–57.

50. BOSC, P., CHAUFFAUT, A. GALIBOURG, M., and HAMON, G. (1986). Une extension de SEQUEL pour permettre l'interrogation floue. *Modèles et Bases de Données*, no. 4. AFCET, Paris, pp. 17–24.

51. DOCKERY, J. (1982). Fuzzy design of military information systems. *Int. J. Man-Machine Studies*, **16**, 1–38.

Index